SOS ... Satellites Down

Robert W. Haeussinger

Some suggest life is predictable; perhaps they have a point, but only to a degree. Without question, constants are important for healthy living. Moreover things we've become accustom to ultimately become ingrained in us. Therein lays the problem.

People born prior to 1980, have no idea how daily life functioned without Wi-Fi, email, text messaging, Twitter, Facebook, selfies, electronic banking or GPS. When those devices no longer work they have no history to fall back on. Likely confusion will become the order of the day.

Society as a whole has now become accustom to those conveniences wherein they become engrained in our lives. Methods previously used before their advent were abandoned or allowed to atrophy to a point they're no longer useable.

*F*riday, May 1, 2020 began like most days in early May. It was a beautiful spring morning in the northern hemisphere. In fact weather throughout much of the world was relatively calm; no storms to speak off; nothing of any consequence. Unbeknownst to anyone, a change of epic portions would have it début; in a manner so minuet no one made note.

*A*t 8:45 that morning Northeast Airline's flight number 515 lifted off the runway at Wold Chamberlain Field in Minneapolis bound for Tokyo; it departed late because there was an issue in the Control Tower. The non-stop flight of the Boeing 787 would take thirteen hours and twenty-two minutes. Upon lift-off the flight headed north northwest up through Canada, then across the southern third of Alaska before heading out over the Bering Sea.

*A*pproximately nineteen minutes inside Alaskan air space a strange phenomenon occurred. The plane's autopilot had been programmed for optimal fuel use. Air speed at the time was five hundred and forty miles an hour at an altitude of thirty-nine thousand feet. Captain Thompson and First Officer Johnson were discussing hotel arrangements upon arrival. Initially Thompson didn't notice the yellow warning light flashing on his console. Johnson however did notice it when he took a map from his flight bag. The light indicated a program change to the plane's course heading. Thompson then rechecked the autopilot settings yet found no change. The light turned off once he acknowledged the indication. However, a few minutes later he sensed the aircraft was turning, ever so slightly. Given the plane's current position he had to wait twelve minutes to communicate with ground control to report the situation.

When Thompson was about to turn his mike on he heard the dispatcher's voice in his headset. The dispatcher said he was tracking their flight on radar, and asked why the plane had switched its program heading. He said the aircraft was now heading off course. Thompson was told to reprogram the autopilot, which he did. Dispatch also said they would continue to track the flight to ensure it remained on course. When they reached Tokyo, maintenance found nothing wrong with either the autopilot or GPS but just to be sure both units were replaced. After that, no one gave it another thought. However, this would not be an isolated incident.

Over the next four days at least five more incidences took place: in China, Russia, France, Brazil, and Norway respectfully. Much like the personnel at Northeast, the other airlines viewed the matter as an isolated anomaly—and didn't share the event with others in the industry. Unbeknownst to anyone, a scientist at NASA's space observatory witnessed unusual activity on the surface of the sun; the pattern and frequency of sunspot activity looked different and appeared abnormal to every previously logged observation. However he concluded the activity was rather minuet and did not believe it was a permanent change from earlier activity; he simply recorded his observation in the log and forgot about it.

For the next three weeks nothing out of the ordinary happened. However on Tuesday May 25, the day after the Memorial Day holiday, weird things reoccurred; this time it affected internet communications, via Wi-Fi. What made it all the more unusual, only three countries were affected, the United States, China, and Russia. In every case, the interruption lasted less than an hour; what did get through during that period was corrupt nonetheless. Making things more confusing, all the interruptions occurred at the exact same time but no one knew that. Speculation was rampant; finger pointing ran the gamut. Heading the list, cyber warfare; nearly every country singled out a perpetrator.

The U.S. accused China and North Korea of masterminding the event. China in turn accused the United States. Russia accused China then Ukraine and lastly the United States. The media also had a field day; needless to say everyone was on edge. Even the U.N. got into the fray, with accusations flying everywhere. The average citizen couldn't figure what all the fuss was about; periodic outages had become common place. On a normal work day, the system could easily become overloaded and things slowed or otherwise stopped altogether; no one was enjoying this. Tempers finally cooled once they— China, Russia, and the U.S.—realized all were equally affected at the same time. Then things became peaceful, at least for a while—then came the afternoon of September 30.

On that occasion, cell phones became the victim. At first, interruptions were limited to drop calls. Later calls were sent to the wrong phone, creating mass confusion. Making it more challenging not everyone was affected; interruptions and errant connections occurred sporadically. Some users never experienced a problem. Almost immediately the blame game resumed, at the top of the list, cell phone carriers followed by the federal government. No one bothered to ask the scientific community what they thought.

No One Wanted to Listen

Once interruptions became well know, NASA and the scientific community began looking into the matter more closely; specifically activity taking place on the surface of the sun. Two of the country's leading scientists; Dr. Mel Jones of MIT and Dr. John Richards of Stanford were selected to assist NASA in the investigation. The White House wanted a preliminary assessment ASAP—an extremely aggressive timeline even in the best of circumstances, which this clearly wasn't. While the scientists went to work, the political pundits and naysayers swung into action; the universally accepted theory among these "experts", cyber warfare. The problem had nothing to do with science, rather cyber terrorists, with the backing of their respective governments or terror groups had plotted to destroy the very framework of how people communicated in the twenty-first century. Moreover, the group felt that without of some sort, the digital network would be rendered useless. Even more troubling than their preposterous theory, they believed the public, not only here but around the world, were accepting their position as gospel. Still, after two weeks of investigating the problem, scientists felt something far more frightening was taking place; something so powerful than no one on earth could control or stop it.

Their initial report was repudiated as pure hogwash; the sun couldn't possibly effect communication, it just wasn't possible. I suppose ignorance is bliss; if something appears too large to comprehend, pretend it doesn't exist, it will then go away. It had to be the work of bad people, nothing more. The federal government even went so far as to threaten NASA's funding if scientists refused to away from such a claim. Even allies of the United States refuted the scientists' claim. A Greek philosopher Plato reportedly said, "Politics is a lie spoken as a truth;" if you repeat something often enough it will be believed. Scientists involved in the investigation weren't deterred despite threats.

The scientists believed sunspots were causing much of the problem, probably because of the frequency at which they were occurring. They also observed new and far more intense activity taking place around the perimeter of the sun; activity not previously observed. Compounding the situation, the intensity seemed to be accelerating at an alarming rate; the long term effects yet unknown. Just watching it made scientists nervous. What did it all mean? Nobody knew. The scientists decided to continue their research in earnest, despite political objections; they would simply maintain a low profile until the political establishment came to their senses.

Verbal Warfare, Among Nations, Approaches Critical Mass

The sun wasn't the only place where explosions were happening; finger pointing and name calling on the ground reached an all time high. Not only was citizenry complaining to government and communication providers; nations were also bickering with each other; at the top of list, the United States and Russia, followed closely by the U.S. and China; U.S. and Europe and Europe and Russia. Threats littered the air waves and dominated print media.

A world once meticulously connected to everyone, electronically that is, was now in disarray. The U.S. threatened to block financial transfers and web site access to countries it thought had a hand in the technical disruptions. Those counties in turn, threatened to do the same to us. People weren't talking to each other about the true cause of the problem. They simply didn't accept the fact or at least the possibility that something external was the cause. The possibility of losing, perhaps permanently a state of the art communication system, both voice and data was unfathomable; it wasn't possible. If the rhetoric continued, hostilities would likely ensue; which it did a month later.

*I*t was Russia who made the first move. Historically, Russia objected to the U.S. Navy's operating in the Mediterranean. Up to now, it was only a minor annoyance. However, when two U.S. destroyers entered the Black Sea, Russia drew the line. Immediately they demanded that the Unites States withdraw its ships and with no delay. With the President's support, the Navy ignored the request. The denial forced Russia to mobilize its air force; an intense standoff ensued. The confrontation so intense, hostilities could have broken out at any moment. The next day, China banned all U.S. airliners from landing in China and directed all State Department officials be sent home.

Everybody blamed everyone else and no one was talking about a solution, not even among themselves. It became so heated, war appeared eminent, and it was just a matter of when and where. Then just as fast things quieted down. Solar activity around the sun had subsided, just as scientists predicted; however they knew it was short lived; for the sake of peace they didn't mention that to anyone outside their inner circle—keeping the peace was the first order of business. In a cloak of secrecy, scientists at NASA developed a disturbance table which predicted when and where communications would be disrupted again and for how long.

What they couldn't predict was how intense the solar disturbance would be and whether the frequency rate would accelerate; there were simply too many variables involved. More importantly there was nothing they could do to stop it. They could only watch. The calm didn't last long; the scientists felt they had to announce what was coming, nearly everyone thought it a sinister plot by some sophisticated terrorists; it couldn't possibly be an act of nature.

On Sunday, May 24, at approximately 11:20 am, electronic communications covering most of the U.S. was cut off completely or operated intermittently. Cell phones were rendered useless, voice and text. Wireless internet, although technically operational, slowed to a crawl; broadcast radio signals were least affected. Three hours later, things were returning to normal, at least in the United States. Much like a weather front, the storm moved east. A short time later nearly every communication device in Europe and western Russia was out of service. Four and a half hours after that, China and most of the Far East was effected. It was then governments of the world realized it was not a terrorist plot but rather a natural event of epic proportions; the scientists were right after all. The problem, most people around the world didn't quite see it that way. Government and big business were the culprits.

As a result of that thinking, a worldwide economic revolution would commence, pitting average people against the large money trusts— i.e., big banks. It was those people who caused this mess. That may have been right, but pointing fingers didn't change reality. In all likelihood the present communication system, consisting of cell phones, satellite transmissions, weather forecasting, banking, and wireless were no longer workable for the foreseeable future, possibly longer. So what was the answer, short and long term? Two generations had grown up with wireless electronics; they knew no other form of communicating. The world was in crisis; leadership was needed, not bickering, finger pointing or accusations.

Citizen Unrest Escalates

Up to then, public displeasures over the communications mix-up was relatively peaceful. Just minor grumbling, letters of concern, and threats to cancel service. However, the frequency of disruptions would grow, effecting not only personal phone calls and messages but financial transactions as well. Individuals had grown accustom to fast and reliable service, now they weren't getting it. Systems they depended on, without question, were letting them down. In the beginning dropped calls and messages happened several times a week for individuals, but then mushroomed to three sometimes four times a day. As a result tensions rose. Calls were not returned; shouting matches between the respective parties unusually ensued after they met in person. To a great many the world had turned upside down right before their eyes, and it scared the hell out of them. There was nowhere to turn for help and most reacted violently, as you might expect.

Since phone lines were jammed or else out of service, customers chose to visit their carrier's place of business to obtain answers, specifically when the problem would be corrected. As you would expect, sites became very crowded, often spilling outdoors. It was pure pandemonium. Fights sprang up everywhere, even ones not directly involved were impacted; a stray punch that failed to hit its target often ended up hitting an innocence bystander. It just wasn't only fists that were swinging kicking was also part of the act. Thankfully, nobody brought a gun; at least not during the first week; that came later. Given the number of clients involved, there simply wasn't enough staff to adequately address customer concerns. Nothing was done to calm their frustration. Brief interludes between altercations mitigated the confusion at least somewhat, but when the interlude ended a much larger confrontations would occurred--all hell had broken loose.

This time, the disruption not only affected cell phone and wireless internet service, everything with the exception of the land lines was down. The bad news, these lines had no additional capacity available to alleviate the situation. Nearly everything ground to halt. No calls, no text, no emails, no money or document transfers. A tragedy of epic proportions had occurred. What did it mean for people's future livelihood? All community and business leaders would say was, "We'll get through this, just be patience." Well people didn't buy it; but then nature reversed course again; the electromagnetic storms stopped and service started to return.

Government and business leaders were more than happy to take credit for the turn around, but scientists knew better. From the very beginning, they said disruption would be sporadic; there would be times, perhaps for several weeks, when everything worked as designed, followed by periods of minimal inconvenience, and then periods when nearly every form of communication was down—i.e., out of service. The scientific community knew the calm would lull people into thinking the worse was over, which it wasn't.

On July 23, Dr. W. Roth from MIT made a startling discovery; the surface of the sun was about to undergo a dramatic transformation; the extent of which not yet clear, nor when it may occur. In a worst case scenario, present wireless communication, because of its design, would become unserviceable; an entirely new system would be needed to replace it. Roth shared his findings with others in the scientific community;

generating both support and skepticism. He was warned, if he shared it with government and business community, he and perhaps the scientific community could be ostracized. Roth wrestled about what to do, if his theory was correct, time was of the essence; it could take several years to design and implement a new system, as such work should begin immediately. What troubled him, what if a seismic event occurred before a new system was ready to take over. It also worried him if he remained quiet and the worst of the worst occurred, was he able to live with that. Given his position at MIT, he was well acquainted with government leaders. It was on that basis, he decided to move forward.

Meeting with Leaders

On August 1, Roth invited his good friend, Senator John Sontag, to take a walk. The two had grown up together, and Roth trusted him completely. In just thirty minutes Roth laid out

22

what he recently discovered as well as providing a suggestion on how to precede. In was decided Sontag would arrange a meeting with the President and the Secretary of State to discuss the situation. The meeting was subsequently scheduled the following week at Camp David Maryland. Roth didn't know the President personally and chatted with Secretary O'Brien on one occasion; he worried the men might reach the wrong conclusion. Neither the President nor Secretary O'Brien knew the true nature of the meeting; instead they relied on Sontag's claim that the meeting was of upmost importance.

The meeting began inauspiciously, like boxers sizing up one another before the fight. Sontag spoke about the need and nature of the meeting then turned things over to Roth. For the next forty-five minutes Roth laid out the very nature of his findings including potential outcomes. The President and never interrupted him while he was talking; they just sat there, appearing awed.

The ensuing dialogue between the President and Roth however got rather testy.

"So Dr. Roth, do you have any idea what you're saying, assuming I believe what I heard and moreover how it may affect the country and the American people?" 'Mr. President I wouldn't be here talking to you if I didn't think it was urgent. If a horrific event takes place, the way we do business and interact with each other will be torn apart. We cannot sir, simply fall back on our previous system; it hasn't the capacity and it certainly is not up to date. I realized people may lose their jobs and businesses might shutter; it could cause an economic disaster. Having said that, I cannot shirk responsibility by simply ignoring the possibility it could happen.' "So Dr. what do you suggest we do, you seem to have all the answers?" 'Sir there is no need to get rude. By inference, you seem to suggest we do nothing. I hope that's not the case. In my opinion, we need to develop short and long range systems that aren't subject to abnormalities in nature. If the event I've outlined occurs, and I believe it will, how will we communicate in a timely manner and will business function normally? People will want answers. May I remind you how chaotic it was during the previous events; they were relatively minor. Do you want to be remembered as the president who sat on his hands and did nothing or someone who stepped up and took action? It will have implication for not only this country but for others around the world, that's why I asked Senator Sontag to have the Secretary of State, join us. Now I've said all I have to say; I think I it's time for me to leave.' The President just sat there and said nothing as Roth walked out. Roth probably thought his career was over, given what he told the President and the manner in which he said it. Although he didn't care, it needed to be said; it was the right thing to do. Senator Sontag soon joined him.

*F*or the next hour, the President sat in his chair and starred out the window without saying a word. Secretary O'Brien tried to engage him in conservation but failed in his attempt. Only later did O'Brien comment, you would've thought war had been declared and maybe it had. As far as anyone knew, President (Bolen) didn't talk to anyone about it for several days. Sontag and O'Brien also said nothing.

*R*oth's drive back to work was a trying one. He went back and forth as to whether speaking to the President was a good idea. The scientist in him said it was; the emotional side thought otherwise. He desperately needed to talk to someone about what he was feeling. He also worried people might take advantage of the situation, not only in the United States but in other parts of the world. Then he thought about how such a disruption would affect his family.

After Effects on Individuals and Companies

*T*o illustrate the affects, Roth tried to imagine the impact on a typical user, i.e. a person who uses a cell phone for personal and business activities, is an active internet user—both wireless and fixed—does most of their transactions, i.e., shopping, banking, etc. on line, and has cable television service. If and when those services are no longer available, perhaps permanently, what else is there? Certainly nothing is convenient or cost effective; and what if they use all of those devices to run their business—things get more complicated.

*T*he impact on businesses would be enormous. Commerce may grind to a halt. Orders would have to be handled the old fashion way--by hand. Given the magnitude of problems, consumers would likely cut back on purchases, especially if their jobs were at stake. The financial community would be devastated as well. Money transactions, such as business orders, would need to be handled another way—minutes would turn into days—transportation costs would soar through the roof. Whether any one company survives could very well depend on if and when a solution, be it temporary, could be found and put in place while a permanent solution is developed.

In a surprise move the President asked Roth to return to the White House. Following that, he wants Roth to meet with the Secretary of Commerce and the Vice-President. Senator Sontag, for some reason, was never invited. The President wasn't about to tip his hand to Congress; he didn't want the issue turned into a political shouting match. Roth had no idea what to expect of the President and had no intention of getting caught in the middle of a political tug of war. He thought about bringing a colleague along—someone he could trust--but in the end thought better of it. If something good could come out of the meeting, he was all for that. As he was about leave, the White House called and said the meeting was canceled; it would not be rescheduled. Roth felt the President was having second thought about pursing the matter. It that was the case, maybe he had to get more involved. If he decided to take that path, he had no idea how to go about it.

Further Consideration by Others

At first he wasn't sure if he should consider alternatives; that might cause more confusion. However, when his daughter came home to visit it changed things. Nicky Roth was a graduate student at Northwestern University studying astrophysics. Later she admitted her visit wasn't a social one. Students on campus were talking about the cell phone debacle, and there was great concern. Facebook, Twitter, Instagram, and other services such as those were very popular with young people; the thought of losing some or all of those services besides voice and messaging gave them pause. They wanted answers but it seemed the only thing they were getting was a run around. Nicky asked her Dad, point blank, what was going on and was anyone doing anything about it. Making things worse, father and daughter were estranged—she was a determined young woman and a free spirit, he far more serious; a nuts and bolts kind of guy.

After a day of pleasantries, Nicky went on the attack; she wanted know what her father knew, more specifically what he intended to do about it. When she brought up the subject he grew defensive, saying he was generally aware of the situation but was too busy with other matters to look into it in great detail. Nicky knew her Dad, far more than he realized or was willing to admit; she saw right through that 'song and dance'. In a highly agitated voice she said, "Dad, cut the crap, I know you know more than that, is it serious?" At first he thought it was in her best interests not to know, but then the scientist in him took over. He said, 'yes, it has the potential for being very serious, with long term consequences.' She wanted to know what that all meant and who or what was causing it.

Roth went on to explain the relationship between activities now occurring on the surface of the sun and the impacts it has on communication transmissions, specifically between satellites in space and the technical devices on earth; devices like cell phones, tablets, computers, and other devices that send and receive information via wireless. The look on her face told the story; things would be different and unless something was done to rectify the situation, if only for a short time, id not chaos would ensue. When she pulled herself together, she asked him what he intends to do. He explained he had spoken to several officials regarding the situation and its long range impact on the country but said he got nowhere. At first officials blamed it on international terrorism; the sun couldn't possibly cause that. Even when the range and frequency of the disruptions increased, they said it was minor and would eventually stop. They simply buried their heads in the sand, hoping it would go away. In a very emotional sounding voice she said, "Dad you can't let this happen, you have to do something, you have to get working on it." She even suggested that her sister and brother in-law get involved—the pair were physics professors at Stanford University. After considerable discussion, she convinced him to fly to California and talk to them. She respectfully declined an offer to accompany him; she wasn't on speaking terms with her brother in-law. Two days later, Roth got boarded a plane and headed west, he had no idea how he would present his case much less tell them what he wanted them to do. He would arrive unannounced, so as not to draw suspicion.

*O*n the morning of November 25, 2020, he awoke following a restless night of sleep; his plane arrived an hour late and by the time he got to the hotel it was already one o'clock in the morning. Following a quick shower and a cup of coffee he decided to call his office to see if anything had changed. However when he switched on his cell phone, there was a message no cell phone owner ever wants to see, "service not available'. When he tried using the hotel phone, he heard, 'all circuits are busy, please try again later'. Failing that, he turned on the TV, same thing, nothing. In the back of his head he knew what was going on but wasn't ready to accept it, not yet. After talking to the desk clerk in the lobby, his fears were confirmed; virtually all cell and wireless communications were down, not only in the United States but also in other parts of the world. His suspicions proved correct; the only thing he didn't know, how long it would last this time. Immediately he hailed a cab and went to visit his daughter; unless something major happened, she would be at work, at least he hoped so. Thankfully the cab driver knew the location of the building where she worked on the Stanford campus.

The Bailey building was a huge twelve story structure located in the southwest quadrant; her office was on the ninth floor. When he walked in, her back was turned so she didn't see him come in. To avoid frightening her, he softly tapped on the door. Her first words, "Dad what are you doing; why didn't you call to tell me you were coming?" His response, 'I think you know the reason.' For the next ninety minutes, he explains as best he could what was taking place and why. She said that was all well and good but why did he come all the way there to tell her. They never collaborated on anything and usually disagree on most everything. The country had a problem, a big problem and needed the best and the brightest to come up with a solution, she and David—his son-in-law—should be part of that group. However, before they could begin working, he had to contact his colleagues back at MIT and get the latest information as to the scope and magnitude of the disruption. He asked to use a dedicated phone line, which was set up by the Defense department for emergency purposes and this certainly was. For the next three hours he took meticulously notes regarding the disturbance.

First the good news, the disruption was expected to end by night fall or at the very latest the following morning, but no guarantee, then the bad news. The event occurred exactly has he predicted; he was only several days off. The magnitude however was far greater than even he imagined. Approximately two-thirds of the world's cell capacity was down, wireless service even more so. GPS and weather satellites were inoperable and so far attempts to reinitialize them had failed. If scenario F1--a scenario he never mentioned to anyone—happens, communication as we know it would disappear for a very long time. Politicians had to be convinced this situation deserved top priority. He equated talking to them as if talking to morons--useless. When he regained his composure, he asked Patty point blank, would she and David consider working with him in developing a short term fix. She said she had to discuss it with him before they could commit. When she left, he put in a call to the White House; he hoped the President would accept the call. The truth was he tried to contact him several times after their meeting was canceled, with no success. Sadly the President didn't pick up now. It would be the first of many long days, not only for him and his family but for others besides. Armageddon or so it seems. Society, as expected, became chaotic; they felt violated and frighten; they needed someone to blame. In Washington, the President decided to call a meeting of his national security team.

A Meeting of Powers That Be

*N*ormally this group would the right one to have a serious discussion, assuming the country's national security was or would be compromised, it had not at least not in the traditional sense. What the hell did a bunch of generals, admirals, and a sprinkling of politicians and bureaucrats know about a dysfunctional communication systems. The President expected the group to give him meaningful advice. Given they didn't understand the nature of the problem and the underlying cause how could they possibly be expected to offer any worthwhile. The meeting last four hours; four hours of bickering and finger pointing; some people went so far as to blame Roth for bring the problem to the President attention in the first place. It was the President seventeen year old daughter who managed to convince her father to work with the scientific community. She told him Dr. Roth understood the problem better than anyone else;

he was the one who had the courage to speak up when others did not. She ended her lecture by telling the President her friends were blaming her for not getting the problem solved, she was the President's daughter after all, if anyone could convince him to get it done it was her. To hear her tell it, losing Facebook, Twitter, cell phone texting and satellite TV was equated to losing a family member. She told him they felt helpless without them. The next day, the President arranged a meeting with NASA scientists and effected vendors and talk options. For some reason, Roth wasn't informed of the meeting nor was he invited; why no one knew.

The meeting was called for the best of reasons; the scientists understood how the sun, specifically certain activity on its surface, affected various communication media. The vendors were there to explain how the disruptions affected their businesses, in the interim and long term. Once the scientists finished explaining the solar phenomenon and its impacts, several vendors accused them of crying wolf; the sun wasn't causing the problem; the activity must be the result of foul play; they wanted government, specifically the President, to get it resolved immediately. As you might expect, the meeting quickly turned into a full scale bitch session. The President let it go far too long, but eventually he put his foot down. In a very agitated voice he says to the vendors, "How is it Dr. Roth accurately predicted these disruptions, I don't call that coincidental?" That shut them up; although one sarcastically said, "So why isn't he here?" It was then the Vice President leaned over and told the President Roth was never invited; truth be told, the President instructed the White House secretary not to call him; why he never said. Now he looked like a man with egg on his face. Subsequently he announced the meeting was over; then he got up from his chair and walked out. Nearly all were stunned by such behavior. When Roth learned of the meeting and what they had discussed he was ambivalent; if they didn't have time for him, he didn't have time either. When the White House tried to call him later he chose not to answer—he was under no obligation. Now there was work to do.

Finding a Solution or Work Around

For the next several weeks he met with his daughter and son-in-law almost every day, two hours in the morning and two hours in the afternoon, all struggling to develop a game plan. Everyone understood that solar phenomena was impacting communication signals; what they didn't know, the nature of the effect; in other words, not just how they were affected but how they were altered in such a fashion to render them useless. Understanding the how, all agreed, was critical to developing a solution. This was the focus of their work, and for the next two months they worked mightily and that very issue. In the meantime, disruptions continued around the world, some lasting several hours, others could last days. The only constant and it was a big one, the problem was not going away; then came a glimmer of hope, be it a small one.

*I*n an attempt to obtain another perspective, Patty without her father's or husband's concurrence asked a second year graduate student to join the discussion on Friday afternoon. The woman accepted the offer, under one condition, she be allowed to bring along her fifteen year old son; she felt the group could use a young person's perspective. The boy's time with the group would be limited to fifteen minutes. During the first ten minutes the boy just listened to the debate then he raised a question. "Were there any kind of signals that weren't affected or otherwise altered?" The others seemed taken aback by the query, what the heck was he talking about, he's just a kid. The silence proved deafening, still he something else to say, "What about AM signals, they're different aren't they?" At that point, his mother said it was time for him to go, which he did. Immediately after he left, the mom apologized. Roth said no apology was necessary, maybe the boy was on to something. So how did a fifteen year old of average intelligence, living in a poor neighborhood come up such an idea? David said, "What the hell is the difference let's look at it, maybe we've been looking at this all wrong."

During each disruption, everyone focused their attention on cell phone and wireless outages, including FM frequencies; no attention was given to AM frequencies. AM media had all but disappeared by 2019, only a couple of dozen stations in the U.S. were in operation. Of the eleven stations contacted, none experienced a signal loss during any of the solar episodes. So what was it about AM signals that made them different? Maybe they were looking for answers in all the wrong places. Even if this new revelation proved fruitless it was still worth looking into. While the three, along with Patty's graduate assistant, continued their investigation social frustration was starting to boil throughout the country and in other parts of the world; it was approaching a breaking boiling point. Demonstrations had become a regular occurrence; although mostly peaceful, some believed it could escalate and turn violent.

Protests Regarding Electronic Injustice

The generation who grew up with cell phones, Wi-Fi, Twitter, Facebook, and other instant media were speaking out about the injustice; they simple didn't know how to communicate with each other in a different way. Direct face to face discussions had virtually disappeared in favor of selfies, texting, and other electronic means. The first major protest occurred in San Francisco; the very heart of Silicon Valley.

*P*lanners, in setting up the event, thought possibly a hundred people would show up; when it started the crowd numbered several dozen. When a camera crew from a local television station arrived, numbers exploded. A short time later, the crowd number in the hundreds. With a bigger audience, the speaker ratcheted up the rhetoric. Loss of service was the chief complaint; the more volatile issue, money.

*I*t seems most, if not all, service providers were unwilling to grant credits to their customers when service was out. That list included satellite, cell phone, and internet providers. Their position, as stated in their service agreements, credits and refunds aren't provided for in the event of a service outage, the rational being a disruption usually only lasts a few minutes but never more than hour. In an attempt to appease crowd, they invited customers to come up on stage—a flat bed truck no less—and talk about their situation.

The first speaker began calmly enough; therein stating his cell phone provider was the largest in the country and that he had been a customer in good standing for ten years. He went on to say, "That over the last thirty-five days, service was unavailable on nine separate occasions, ranging from several hours to several days. He also said he contacted customer service regarding the problem and was told they were working on it. When he asked a about a credit for the time his phone wasn't in service, they just brushed him off, saying their policy did not offer credits. Now the company no longer answers the phone. His cell is useless without service, yet the company expects him to pay despite the outage." On that point, the crowd erupts yelling "service, service, service", at the top of their lungs. As you might expect pushing and shoving ensued. The next speaker was even more vocal.

This fellow began by holding up his internet and satellite television bills for the last two months.

In an extremely agitated voice he shouts, "In my hand I have over three hundred dollars in bills for service unavailable fifty percent of the time. If you ordered four pizzas and they delivered two but billed you for four you'd be pissed, right. Will I'm pissed and the sons-of-bitches won't do a damm thing about it. They say it's not their problem; well who the hell's problem is it, it's not mine though they expect me to pay as if it was. To make matters worse, they actually expect me to pay a cancellation fee if I break ties with them. I say we can't sit back and just take it; now that's all I have to say." The crowd erupts as before and but this time its profanities their yelling. Nearly everyone there is highly educated and earning a generous income but given their present mood they're acting more like a mob about to riot. The crowd organizer, sensing things were getting out of hand tried to calm everyone.

*A*s before, he started enough. He told them it was time to use their head not their brawn. Moreover he said they had to put collective pressure on the companies involved while demanding that government protect consumers from outlandish practices besides supporting the scientific community's efforts to find a lasting alternative to the situation. For a moment it seemed the message we getting through, but even that was short lived. There was plenty of bent up anger; that energy had to go somewhere.

*W*hen dozens of squad cars showed up all hell broke loose; the cops were caught off guard. As they approached the ball of humanity, the crowd reacted instinctively by pushing officers back and in the process knocked many of them down. Several squad cars were even accosted and/or turned over. It became a riot of major proportions, reminiscent of 1967 Detroit riots, but this time it wasn't about race, and it was about greed. Eventually fire trucks arrived and dispersed the crowd with hoses. In total over fifty people were arrested including the event's organizers. Over the next three weeks similar protests occurred in other parts of the country. The knuckleheads in Washington didn't have a clue about what was going on much less knowing how to stop it. Senator Sontag said we needed to get things moving. It was imperative they have face to face talk with the President. After much badgering, Bolen agreed to a meeting. Sontag insisted Dr. Roth join them.

The meeting was held in the Vice President's office—the VP was out of town—not even the President's Chief of Staff or his personal secretary knew about the meeting—it was that hush, hush. Sontag spoke first. He told the President things had gotten way out of hand; we simply can't sit back and do nothing. If further disruption in communication continues, things will get much worse. At that point Dr. Roth tried to explain, as best he could, the very nature of the solar phenomenon and where we find ourselves currently.

Solar flare-ups are not unusual in fact they typically occur every eleven years or so. The magnitude of these events is usually quite small, amounting to minor inconveniences ranging from a few hours to a few days. However, this time it's different; solar flares are now much stronger; greater than anything seen in modern times. The last time an event that approached this magnitude occurred back in 1859. Electrical systems were in their infancy, still the damage was enormous. I suspect, although I hope I'm wrong, a cataclysmic event could very well strike, and if it does, it could set us back at least three decades. Industry and commerce will be greatly affected, resulting in massive layoffs. It could very well spark a worldwide depression and/or a revolution, assuming its impacts are spread equally, which is possible. He went on to say, he wished he had a solution but he does not, no one does. Right now he and several family members are trying to come up with answers. One, how to get us through in the short term, what alternatives are available, and two in the event of a cataclysmic event, what can we come up with to address our long term needs. The President responded by asking him what the government had to do. At that point Senator Sontag reentered the discussion.

Sontag said the first thing government has to do is rein in the communication industry. Refusing to provide billing credits for service not otherwise provided alienates the public. The incidence in San Francisco is just the tip of the iceberg; will it hurt their bottom line, absolutely. Destroying buildings and equipment; injuring or perhaps killing others would be far worse. Sound leadership is critical. All the resources presently available must be put to maximum use, not next week, not next month but right now; anything short of maximum effort on the government's part would be a travesty. There is simply no time to waste. We are at war and our enemy, time, not only our lives but the future of others is at stake. The President seemed traumatized by what he was told; Roth and Sontag didn't know what to think. They said what had to be said, there was no getting around it; the facts speak for themselves. They needed a break. As both men left the room they couldn't help shaking their heads. They could care less if the President tried to discredit either of them; in the end he would be the one responsible. Leaders of industry and commerce were under the delusion it would all go away; it was the scientists who were stirring things up, if anyone was to blame it's them. Up to this point the media seem to concur. That would change once a well know Wall Street publication released a blistering editorial on the subject. What follows is an excerpt of their article.

*"I*t has come to our attentions the bastions of industry; along with their counterpart in government seem to deny the existence of solar disturbances affecting us here in the United States and in other parts of the world. These disturbances are troubling. Up to now, we have developed and presently operate the finest communication system ever. In a matter of seconds signals can be sent to nearly every place in the world; be that voice, video, audio, or text. In fact, the system has been so reliable and perhaps unlimited regarding its full potential we have chosen to abandon previous systems. We know that Earth was affected by solar storms before, and most of our lifetime be they more minor events. That was not the case in 1859, then our electric system was in the Stone Age compared to today, but its affect was nonetheless devastating. Our friends at NASA and MIT tell us these disturbances are now occurring on the surface of the sun is far stronger than those back in the 19th century. The simple fact is we have no

backup system in place to handle our communication needs, should a catastrophe occur. Can we say, without a doubt, such a cataclysmic event will happen or when, the answer is probably no. But we've been around long enough to know you don't wait until it happens, and it appears we are doing just that right now, with the possible exception of a couple of people. Should the prospect of such an event scare the hell out of us, you're darn right it should. Our high speed economic will grind to a halt, businesses will fail; people will lose their jobs, their homes, their savings; we can't let that happen, we needed to put every available resource we have towards handling this...

Around the World Reaction

Reaction around the world seemed mixed. Generally our allies thought there was a chance possibility a problem could arise but they didn't appeared too concerned about it, up to that point disruptions in their part of the world were far less than anything happening in the US. Without saying as much, they seem to take a 'wait and see' position. There appears little hope of changing their mind, absent a crisis. Enemies and political critics alike have a different take.

Those folks think or at the least claim, the U. S. alone caused the problem. They dispelled the idea it's an act of nature, despite evidence to the contrary. They simply view it as another attempt to control everything and everyone. Americans are bullies by nature, or so they believe. People, like the politicians around the country, have their head in the sand; 'if you close your eyes you can't see a problem, therefore there is none, end of discussion'. The irony of all this, a goodly number of scientists from all parts of the world admit something very disturbing is taking place on the surface of the sun, it demands attention.

Absent the government, citizens from other parts of the world expressed those concerns although they aren't vocal about it. Maybe part of the reason, they weren't tech dependent; thus no need for instance gratification, via electronic devices. However, there is no hard evidence to support that conclusion. The media, not only in the U.S. but also around the world, are more interested in dispensing hype than hard facts. The blame game sells newspapers and boosts ratings, offering a solution does little for them.

Weather in much of the world during the months of October and November was rather calm. No storms of any magnitude, precipitation; and temperatures virtually identical to the historic average. More importantly there were no electrical disruptions anywhere. These facts enabled us and others to become complacent. On December 1, at 8:42 AM eastern standard time, complacency came to an abrupt halt.

Nature Speaks Round ONE

Within minutes massive electrical storms
erupted on the surface of the sun, much greater
than anything previously seen. At 9:05 a
colleague at MIT put in a call to Dr. Roth; at
that time he was still in California. Immediately
Roth hoped a plane home. Minutes after he
arrived, he knew the country was in trouble—he
earlier predicted a series of large, possibly
catastrophic events. In front of him were video
feeds from several high power telescopes aimed
at the surface of the sun. In every instance, solar
flares were shouting out from the surface, some
as far as two miles. Other devices in the lab
recorded electromagnetic levels virtually off the
chart. If levels of this magnitude lasted more
than an hour, most satellites presently in orbit
would be impacted in some way; even
permanently damaged. The stakes were huge;
leaders from around the world must be notified.
Roth asked his administrative assistant to call

the White House. Fortunately the White House had a call into him. It had taken twenty minutes to locate him.

The President told Roth he heard reports regarding a possible event but no one could tell him about its severity or its impact. Roth didn't beat around the bush, the event would likely cause massive damage; he had no way of predicting how extensive it might be but offered an educated guess. In order to lessen the impact, major steps had to be taken immediately, the next ninety minutes were crucial. However before he could continue, the President interrupts him, asking what areas would most likely be affected.

Roth told the President, assuming the level of activity remains unchanged for thirty minutes or more, a high percentage of the satellites now in orbit around the world will either be damaged or operationally destroyed, including the

International Space Station. He went on to say some of the satellites might be spared if they're shut down soon enough, before the electromagnetic field reaches them. Virtually every satellite placed in orbit since 2007 have mechanisms onboard to protect them in the event of electromagnetic storms, but it's only effective if the satellite is operationally shut down. What we didn't know, how much voltage each could withstand and for how long. Shut down procedures, based on what he was told takes fifteen minutes to complete. Once a satellite is shut down, systems dependent on that device are also out of service—wireless, cell, television, radio, GPS. From a standpoint of life and death, aircraft are most at risk. Navigation, radar, radio, and similar devices will go down as well. A decision would have to be made shortly about whether to ground all aircraft presently in the air; those awaiting take off would be told to abort. Those in the air would be ordered to divert to the nearest airport capable of accommodating them, a manner similarly used on September 11, 2001. The President had to decide a course of action. If the decision was to act, and the solar storm proved less intensive, the President would surely take the heat. If he decided to hold off, and the impact was similar to what Roth predicted the consequences would be huge, people could die. The President asked Roth what he should do. Roth's only comment, "when people's lives are at stake, error on the side of caution; for now I'll continue to monitor the storm and provide you updates as needed." It took the President nearly ten minutes to decide. Meanwhile, the media got wind of the story, and you can image the chaos that results.

If you listened to those bubble-head reporters, it was gloom and doom. Nature was going to throw us back in time; before cell phones, wireless internet, satellite TV, instant ordering and messaging, airline service, etc. etc. Topping that, businesses would fail in record numbers; people would lose their job and means of supporting themselves and their family. What they failed to mentioned, the event was actually a wake-up call, that life and nature changes in time, and to survive one must adjust to those changes even if it is forced on them through no fault of their own. Without question, pandemonium ruled. The rule of order was going to be tested and tested severely. In Washington, the President made his decision. He asked the Vice President to assist him in implementing his directives.

Step1, all U.S. satellites, be they owned by the government or private business be immediately shut down and remain so until informed otherwise. Steps 2, all aircraft presently on US soil are hereby grounded and shall remain grounded until told otherwise. All aircraft presently flying in U.S. air space or otherwise destine to a U.S. airport are hereby ordered to land at the nearest available airport capable of handling their type of aircraft and upon landing shall remain grounded unless told otherwise. Step 3, all public and private electrical generating facilities presently operating in the United States, are ordered to institute Paragraph II of Emergency Diasaster Plan (B)aker, dated September 11, 2001. Step 4, all radio and television stations licensed to operate in the United States, are hereby instructed to implement Paragraph III of Emergency Diasaster Plan (B)aker, dated September 30, 2011. Paragraph III limits the scope and nature of programming in the event a emergency disaster is declared—primarily news and weather information only. Step 5, the Secretary of the Army shall oversee and coordinate the use of the Emergency Broadcast System at the direction of the Vice President. All directives became effective as of 9:49 AM eastern daylight time. At 9:55 the President would address the American people, he was not looking forward to it. His political future and his place in history would be defined on how well he handled things. What follows is a summary of those remarks.

"My fellow Americans, I come before you this morning with grave news. By now, some of you have heard about the situation we're about to face. Nothing of this magnitude has been presented itself that might permanently impact our daily lives. The truth is we don't know the full implications of this phenemon, the best we can do is make some educated guesses and error on the side of caution, especially when human life is a stake. Whatever the outcome of this and future events prove to be, our way of life will no doubt change, there is no denying that. We can either cower or start exploring new possibilities; I choose the later and I hope you will to. I shall keep you informed of new developments as they become know. I ask for your cooperation and patience during this difficult time; together we'll get through it."

The procedures the President outlined in his address were good ones; the trouble was they were instituted too late. A mere eighteen minutes later, the first round of electromagnetic shock waves penetrated the atmosphere. For the first few minutes, disruption was minimal, and then all hell broke loose. One right after another every U.S. public and private satellite, except one, failed—whether the failure was permanent wouldn't be known for days. Eventual the last functioning unit failed not because of the electrical storm, but from a circuit overload. Before it went down, there were reports that other countries experienced similar problems. From then on, problems multiplied. Airliners, military aircraft, and private planes were having problems with their navigation systems; ground radar used to track airborne traffic proved unreliable, often providing errant information. Planes in many cases flew blindly. Fortunately most of the aircraft flying inside U.S. airspace had already made contact with a divert field and were heading there, with a little luck most would be on the ground inside an hour. Also working in our favor, the FAA had a plan in place to recover (land) said aircraft absent radar or voice transmission. Light signals were used to communicate specific landing instructions. Based on the directive pilots were required to carry; a specific flight pattern was to be maintained until instructed to land. It was a rudimentary system but it worked for the most part. Airplanes weren't the only problem; electronic banking and stock trading was also in a quagmire.

In this country millions of dollars are transferred each minute; on trading days, hundreds of thousands of stocks are bought and sold, amounting to millions of dollars. When the electromagnetic shock wave hit earth, thousands of transactions were in process. Some would be lost, digitally speaking, however the vast majority transmitted to the wrong place and no subsequent trail made of that location. It could

take years to locate those transactions; the losers would likely outnumbered winners. Life savings and business assets perhaps wiped out.
Airplanes and financial activity weren't the only problems; the electrical grid was facing its biggest challenge.

In 1859, when the last solar storm caused major damage, the use of electricity was quite limited, aka, the telegraph—it was the Stone Age of technology compared to today's systems. Nevertheless the 59 storm took out most of that system down; perhaps destroyed would be a more appropriate word. Not long ago, generating capacity was localized, covering maybe one or possibly two cities or perhaps several counties. Today, a community in Kansas might get most of its power from a plant in Wisconsin. Electrically speaking we'd become an interdependent country. 1859 looked primitive by comparison.

*F*ifteen minutes following the initial shock wave, more than seventy percent of the grid was down; numerous communities along with certain regions in the country were without power. No power to operate businesses, no power to heat homes, operate a refrigerator, pump gas for ones car and lots of other things. Most of the old power plants were shuttered; the ones still operating were only able to provide a small fraction of our electrical needs. Earlier, cell phone, Wi-Fi, and Twitter disruptions by comparison, were minor annoyances. This time the storm got everyone's attention and they were frightened because of it. By 1:30 PM eastern daylight time, the storm ended; by then the damage was done. Life wouldn't be the same; the public looked to government for answers and help to maintain order. Still, there was a question if they were up to the task on either account. Was the Wild West about to return?

Government Responds - Possibly

*L*eaders in Washington were just as confused about what to do as was everyone else. Compounding the situation, other parts of the world were equally affected, with the exception of Russia; they were barely affected; at least that's what we thought. The President was in shock, and seemed incapable of making rational decisions. It was Vice President Webber who stepped up and took charge, much to the Cabinet's and Congress's displeasure. Congress also seemed reluctant to declare the President incapacitated. Fortunately the President decided to transfer power to the Vice President just a few days later, at least until he recovered from his malady. Webber almost immediately ordered a thorough assessment of the situation—it proved a daunting task. That assessment included maintaining a watchful eye on our enemies; they may try to take advantage of the situation.

Although the Vice President was invited to attend the meeting the President hastily arranged with Dr. Roth, she couldn't; she had a prior commitment, thus declined. Nevertheless, she heard rumors concerning the possible threat; which were proved to be correct. Unfortunately, she incorrectly assumed they were made up by some crackpot academic or some other crazy who disagreed with how the government spends its money. Still her gut told her to look into it further to know if they were any truth to it; she asked Senator Sontag to contact Dr. Roth for help. Unfortunately the Vice President was not held in high inside the academic community; as member of Congress she repeatedly introduced proposals to reduce or eliminate funding for academic research. It took the Senator nearly an hour to convince Roth to work with her. At that point, Roth was not aware of the President condition or that he temporarily transferred power to the VP. Webber, Sontag, and Roth meet the following day in the Vice-President's office. When the men entered the room they quickly sensed the tension—the Vice President's expression said it all.

Once chilly handshakes were over, the Vice President got right to the point, what was Dr. Roth's understanding both before and after the storm? At first, Roth said nothing; he felt the tone of the question quite rude, suggesting he was, to blame in some way; only after Sontag's repeated insistence, did he decide to open up. In a calm but slightly agitated voice, he told her what he told the President when he met with him the first time and what it meant, not only for this country but for the rest of the world as well. Despite the sporadic and petty interruptions, he made his recommendations know, although most were ignored. He said if the steps he identified had been taken immediately instead of waiting; things would be different; when they were implemented it was already too late. Now we find ourselves with real problems. He also mentions that additional storms are quite probable, even near the magnitude just experienced. He said the type of phenomena we're seeing is quite new, so we don't fully understand it as of yet. Since the middle of the 1850's scientists knew a storm of this magnitude was possible one day, but they had no idea when. By ignoring the possibility and not taking it into account the reality of that possibility is now readily apparent. Roth ended by saying, "We have in place measures to protect us from weapons of mass destruction, but we've chosen not to protect ourselves from the force of nature. Our communication and electrical distribution network is the finest man could hope to design, but great as it is/was, the system is fragile and venerable to forces we don't fully understand." As Roth spoke the Vice President sat there in disbelief. Sensing her desperation, Sontag tells her Roth, his family and his colleagues at MIT are working on finding a short term resolution to get systems operating again. Roth quickly responded by saying nothing has proven effective thus far, but is hopeful. The VP's only response, "We need to have a follow-up meeting. She was not looking forward to reading the assessment report that just arrived.

The Aftermath and What's Left

On the morning of December 24th, Christmas Eve no less, Webber was briefed on the current state of the nation. She assumed the worst and her assumption was correct. Preliminary reports indicate nearly two-thirds of the nation's power grid was down, some beyond repair. All but one satellite were destroyed or severely damaged. Cell phone capacity most everywhere was none existent. Initial inspection of primary cell towers did not look good. Engineers say cell communication, voice and text will not return in the near term—they didn't define how long. On a positive note, airline service could possibly resume, be it on limited scale, once ground radar and GPS tracking systems are rerouted to the only operational satellite. Businesses and communities also in a dilemma, without the ability to place orders they cannot service needs. Peoples' ability to get to work is also a problem because gas station pumps aren't working. In some cases they are told to stay home because there was no work for them. Since the power grid is down, people cannot heat their home, nor operate their refrigerator; food is rapidly spoiling. Police department were now having a rough time maintaining order; they tell us pandemonium could erupt at any time. People want to know who is to blame, government, another country or some radical group. Military readiness however is far less precarious.

In 1991 Pentagon officials had the foresight to develop an alternative communication system within the armed forces network in the event the traditional system failed. The only service unit impacted to any great degree is the Air Force. Ground and Naval forces were still able to communicate, be it a bit slower. That fact was especially critical in the Far East. Call it a freak of nature but our enemy, North Korea, was hit by the storm as well; but to what extend their systems were impacted remains a mystery; we must assume it was minimal. As a precaution, the Pentagon ordered the 6th fleet to the patrol waters in the Sea of Japan, north and parallel to the North Korean border. Should the North decide to launch a strike, our ships are in a position to block the assault. Nevertheless, intelligence coming out of South Korea suggested the North was up to something. If true, we can't take any chances. Far worse was the financial community, they were dire straits.

Stock and investment trading was at standstill; billions of dollars now in limbo. Banks as well as investment firms were worried, losses could be enormous also there is a high probability some firms would fail as a result. They want something done and done now. Resolve however was nearly impossible in the short term. Still that didn't prevent them from hounding Washington, principally Congress; although Webber, as acting President got her share to. Merchants from around the country were echoing much of what Wall Street was saying. Their wrath however was directed at local officials and their respective governors. Rioting and looting topped their fears. In Washington, Webber struggled with developing a plan much less a strategy.

A Preliminary Course of Action

*I*n the three years she was Vice President, Webber was not a part of the decision making process; and she made it clear to the President how she felt about that. Looking back, you might say she burned a few bridges; nearly everyone in the White House found her contentious and hard to deal with. With a reputation like that, it's hard to build a coalition even in the best of times, which they aren't right now. As much as she distained academic people, she knew Roth could get things done. As such, she invited him, his daughter Patty, his son-in-law David, the electronic expert, and several of his colleagues from MIT to join her at the White House; the purpose talk strategy. They met three days later.

Roth quickly proposed a two prong approach. First, determine if any of the satellites are repairable despite our limited ability to get to them. If there are three or more, and we have the parts, we could possibly deploy one of our remaining shuttles to facilitate repairs (NASA phased out the shuttle program nearly twenty years ago, however two were kept in operational readiness). Flights would have to be arranged quickly with little time for advanced planning. In conjunction with that, we need to accelerate the launch of four newly built satellites. Under the best of circumstances, it would take ten hours to configure each unit before they became fully functional in their assigned orbit. Second, utilize a highly speculative procedure using our existing towers to provide point to point (i.e., ground to ground) communication using an experimental broadcast frequency just out of the lab. Early tests show electromagnetic waves, likes those emitted from the sun, have almost no effect on that frequency. To do this, it would take four weeks; two for extensive testing, another two for programming. If additional storms occur, only the last suggestion would prove workable. As Roth summarized his remarks, Webber seemed concerned about the enormity of it all; and more so about whether it would work.

*F*or five grueling hours, the group debated the merits of each idea. Just when it seemed consensus was near, someone raised another concern and debate continued. During the discussion, Roth purposely remained silent, not wishing to influence the decision one way or another; it had to be their decision. Although in the end, Webber had the final say; she controlled the resources. While debate raged in Washington, a much larger drama played out across the country; families were barely getting by including those generally well off, the problem, universal, its effects were not.

Impact on Families of Varying Means

Consider for a moment the plight of a middle class family living in the Midwest; let's call them the Allen's. To make ends meet, husband and wife work fulltime to support themselves and their three kids; age nine, twelve, and fourteen. Take home pay after taxes and insurance, combined is $1,200 a week. However, after paying the mortgage, household expenses, etc. a little over $100 is left. The children, like most kids their age enjoy various niceties: laptops, tablets, cell phones, not to mention satellite TV. Paychecks are also deposited electronically. Like most families in 2020/2021 the Allen's are electronically dependent. All of that changed in an instant; its affect horrific.

*P*rior to the mega storm hitting the U. S., the sporadic disruptions that occurred earlier were, by comparison, minor annoyances; within a day maybe two nearly everything was back to normal. At first people thought this storm would be much the same, then two days went by, then a week, and nothing came back. Cell phones and cell messaging unavailable, internet service virtually absent and the part that did work operated at speeds similar to a phone modem. Shopping became a nightmare; credit cards could not be validated—hence no sale; a short time later most stores stopped using manual imprinters to record charge card sales; shortly after that checks were no longer accepted, due to validation concerns; cash was the only option. Since ATMs use credit or ATM cards to dispense cash that option wasn't available. Families, by necessity scaled back purchases; saving what

little money—cash-- they for had for essentials,
like food. It didn't take long before the crisis
affected jobs.

When people went to work a week or two later,
employers they were told business was down, as
such they had to lay off personnel, usually then
and there; spouses were not immune. Sadly lots
of folks had no rainy day fund to fall back on or
if they had one, it might last a few weeks; maybe
a month if they were lucky; there was also a
question about unemployment insurance;
presently banks weren't able to make transfers;
there was no way funds could be deposited into
account(s). It didn't take long for panic to set in.

Kids were often the first to react. They couldn't
understand why this was happening. They felt an
overwhelming need to blame someone, anyone,
but how does one get even with nature; instead
they took their frustrations out on others. Often
it started with a slight or a dig directed towards

another, like commenting on what was said or done they didn't like, and if that weren't enough, pushing and/or shoving might result. It didn't take long for fights to ensue. As you might image, adults got into the act. Bitching about petty or frivolous matters became the new norm; in some cases it escalated into a shouting match. Feuds between neighbors became an everyday occurrence. Families, with modest incomes or well to do were the same boat. Peaceful neighborhoods turned into urban war zones; no area was immune.

Folks earning six figures had much the same problems, be it on a larger scale. Consider the Smith family. Joe was vice president for a major manufacturing firm; his annual salary $300,000; his wife Rebecca a tenured college professor earned $135,000. One child lived at home; the others married with a family of their own. Rick was nineteen and was a fulltime student at a local university. Yet even with all that income, the family had plenty of expenses. Their house, a mansion of sorts, carried with it a huge mortgage. They also had two vacation homes-- each mortgaged--a fancy forty foot cruiser, three imported cars, and a whole host of electronic devices. Financially they had little in reserve to tide them over should disaster strike.

Even before the storm, Joe's company was experiencing a downturn, compounding that, its corporate stock was at a five year low. Making things worse, sales forecasts for the next twelve months weren't dismal at best. Once the storm hit, the company's ability to track sales, order components used in production or manage assets was gone. Suppliers and financial institutions they owed money to demanded immediate payment. Since production had to be curtailed, workers were sent home; there was nothing for them to do nor was there any money to pay them. By the end of the month, the firm's president said several people in upper

management would likely be let go. The very
next day, Joe got his notice; they didn't have the
decency to tell him in person. Given the firm's
limited resources, there was no severance
allowance. Joe was beside himself, the family
couldn't live on Rebecca's salary alone.
Arguments quickly ensued.

*J*oe started drinking then sleeping in late.
Rebecca often blamed him for their plight; why
didn't he see this coming and why did they need
the best of the best. Shouting sessions typically
ended with one or both throwing something.
Rick, who was quite spoiled couldn't understand
why all of this was happening to him; it wasn't
his fault so why should he have to suffer.
Repeated attempts to sell some of their 'toys' and
reduce liabilities went nowhere; people that
showed interest offered far less than what the
item was truly worth. Social contacts also dried
up; folks probably didn't want to be around
people struggling, less they find themselves in the

same situation. The Smith's needed hope, but none was forth coming. Besides families, businesses, local banks owned by small investors, others to suffered a similar fate.

Consider the predicament Ally Bank of Appleton, Wisconsin found itself in. Ally was founded in 1940 by a small group of farmers. During the war business was slow but they managed to get by. By the mid fifties, business started to take off and the bank diversified into other areas outside of agriculture. In 2015, in addition to its primary location, it had six branch offices. Assets at that time were close to three quarters of billion based on the latest audit. In 2018, part of the bank's assets were sold to a private equity firm—most of the long time owners were well passed retirement age, and thought it was the right time to sell. Fantastic customer service, that helped the bank get where it was, began to deteriorate. The bank also began assuming greater risk, to achieve higher profit margins. Loan failure, virtually nonexistent before, crept upward, so slow nobody paid any attention. When staff was let go after the storm hit, other obligations didn't disappear. And since most of the electronic trading network was out of service, the bank had a tough time controlling and otherwise managing its assets. Negative cash flow became a real possibility; there was nothing left to fall back on.

Individuals, families and companies demanded solutions, immediate and long term. The public grew skeptical of government's effectiveness—local, state and federal—specifically their ability to get things done, much less in a timely manner. Unfortunately they were the only act in town. Webber, the acting President, was still trying to finding her way. Blight wasn't confined to small groups of people or organizations in certain areas. It was wide spread across the country; but especially so in the northern states.

Will Government Get Going

*S*ince Dr. Roth and Webber last met nothing had ensued with regard to his recommendations. It seems she relegated them to committees, but the country had no time for committees. So why all the stalling, what was the game? Fortunately she started feeling the heat so she asked for Roth's help once again. He made it clear he was tired of talking; they needed to get things going. She assured him they would be.

*F*or the past few weeks, Roth and his daughter Patty were conducting experiments using Amplitude Modulated (AM for short) waves to determine if they could be modified or transmitted in a different way so as to restore at least part of the communication network. In conducting those experiments they introduced high concentrations of electromagnetic shock waves into the mix, similar to those the sun was emitting, in an attempt to squash the notion that

AM waves were any kind of solution. The results they obtained were surprising, but would it work on a larger scale and would the government be willing fund development. Together they felt confident enough to present their findings. Webber was not aware of their work much less their findings; Roth caught her by surprise.

The meeting at the White House took a serious tone right from the beginning. Commerce Secretary Campbell, who was invited as well, gave an overview of things as they stood. Based on the latest reports, only one of our satellites is operating right now. The status of the back-up units, those earmarked for emergency use, is not known; efforts to get them operational have so far failed. On the bright side, nearly twenty percent of our GPS units are working to some degree, but their response times are extremely slow; estimates, less than one third.

The electrical power grid, like our satellites, sustained heavy damage as well. Presently there are electrical outages throughout the country; from Chicago to New York to California; electrical generation a mere twenty percent of peak capacity. Out west the situation is much the same; from Washington to the southern tip of California, output a mere thirty percent of capacity. The Midwest and the south, including Texas are trying to make up the deficit, but quickly get overloaded from high demand.

Shortage of electrical power is making it harder for communities to operate their utilities. Often there are long interruptions in pumping water and/or treating sewage—well pumps need electricity to operate; sewer plants need electricity to process sewage inflows. Getting basic necessities, like food and gasoline is also proving difficult. Police, fire, and ambulance services as well are often unable to respond to calls for help. People are angry.

*F*riends and family normally to each other now argue frequently. Also there's a lot of screaming going on. It takes little provocation for fights to break out. People in many cases have taken to carrying weapons, either for protection or to influence a dispute. Rioting and looting, at least for now, is rare although it could escalate any time, especially if the situation continues. Order is breaking down.

*S*ome in the room thought the Secretary's assessment were harsh and overblown. Skeptics quickly changed their mind when the Secretary played a video taken the day before. Everyone was greatly affected by it, including Acting President Webber. It was decided they needed a break. When Roth walked into the hall, a White House staffer mentioned he had an important phone call. Roth had no idea who it was.

*O*n the line was Dr. Robert Davis, head scientist for NASA. The news he had was not good. Based on the latest observations from the surface of the sun, yet another electromagnetic storm in the offing; based on earlier observations, this one could strike inside a week, but that was just a guess, it could happen sooner, maybe in a couple of days. Davis concluded by saying this one looks far bigger. Roth dreaded mentioning it to the others. At least there was time to implement certain measures, hopefully lessening the impact. He was the last one to return to the meeting; they sensed something was wrong, given the expression on his face; perhaps something horrible. He asked the Commerce Secretary to relinquish the floor to make an announcement. He struggled hard to maintain his composure.

"Ladies and gentlemen I just received word from Dr. Robert Davis, head scientist at NASA, that another electromagnetic storm will likely strike Earth, probably within the next four or five days. Projections also suggest smaller events will occur in advance of the major event, maybe within a day, perhaps two, we just don't know, as to how many, we're uncertain. In any case there is time to implement several measures. However, the authority to implement them must come from the White House, specifically from the lady over there, Acting President Webber."

Webber read Roth's measures just once, and briefly at that. In a flash, everyone turned their attention to her for some kind of response, but none was forth coming. It was major decision that had to be made, she just wasn't sure she was prepared to make it. Everything was happening far too fast, she needed time to think but there wasn't any. With a dazed look, quite apparent to everyone, she responds, "Prepare the paper work for my signature"; then she walks out. Roth didn't know what to think. Was she the right person to get us through this critical juncture? He had doubts, as did others. However, Roth and the VP had several things in common; each lost a spouse close to the same time and both were close to the same age. Still, he knew, after hearing from others, she was a hard one to figure out and then there was that standoffish personality. She only agreed to run for Vice President out of party loyalty, nothing more. Now, she was in charge, just not as accountable as the President. In any case, Roth wanted to speak to her in alone. The way he went about it was hardly diplomatic.

When Roth left the room he immediately spotted her chatting with a White House staffer just down the hall. In a bold manner, he promptly interrupts the conservation saying only that he needs to speak with her right away. Once the staffer is out of ear shot, he looks directly at her and says, "You and I need to talk, in private." Needless to say, she had a stunned look on her face; no one ever spoke to her quite that way, except maybe her brother but certainly not her husband. When she failed to speak up concern his demand, he pipes up "Where can we go and not be disturbed?" In a stern reply, she says, "follow me." She chose a small office which directly abutted the Situation Room. Once the door was closed, she went off on him.

Who was he to talk to her in that tone, it was rude and highly inappropriate. At first, Roth apologized as to his choice of words and the abruptness he used, but that's where the apology ended. Roth reminded her the country was in hell of fix, people's lives had turned upside down and all Washington could say was, 'We'll look into it.' "You people can't wish the problem away simply by ignoring it or pretending it never happened. People in positions of authority have to step up, that's your job madam Vice President given the President's present circumstance. If by saying that disturbs you, I'm glad. We have to work together, me on the scientific side, you and those fools in Congress on the political front. I don't have all the answers, hell I have damm few, but I do have lots of ideas. So madam, are we going to join hands or butt heads." For the longest time, they just starred at one another without either saying anything. After what seemed like an eternity—more like a several minutes—she says 'thank you.' "You're welcome." Suddenly, in the emotion of the moment, each embrace. Without knowing it, a bond had formed, much to their surprise. When they left, they went their separate ways, though linked by a common purpose; getting the country back on its feet. Webber managed to convince Congress to appropriate the needed funds to develop short and long range programs for restoring electrical and communication systems all across the country. However, before the pending storm strikes, there was a bigger problem government had to address.

Military Mobilization, Threat or Reality

Naval intelligence reported that China and their ally North Korea were putting together a massive exercise, one that all the markings of a major military strike. For quite a long time, the U.S. suspected the Koreans might invade the land to the south (that being South Korea) as well as Japan but what was China up to. Over the last several years, relations between the tows had been good, despite petty disagreements, particularly over cyber theft. U.S. relations with Russia, on the other hand had been going downhill for some time. Was China now siding with our enemy to the north and were they intending to engage Russia in their scheme, the two were usually at odds with each another. In any case, the U.S. was not in a position to check hostilities, wherever came. The only hope for diffusing this situation was diplomacy. The State Department went to work to find out what was going on, with help from the CIA. The last thing they wanted to do was frighten the American people about a possible attack on the homeland, they had other things to worry about.

When earlier and far smaller electromagnetic storms struck Earth, much of Asia and a large portion of Russia was virtually untouched, just a few or rather minor interruptions occurred, nothing major. Even when the last storm hit North America, a large portion of Europe experienced minimal damage. This time the scientific community predicted those areas wouldn't be as lucky; the impact here far less by comparison. Intelligence also indicated China, Russia, and North Korea weren't taking precautions should a storm strike. Their attitude seemed one of indifference. As a humanitarian gesture our scientists tried to warn them, but mostly their words fell on deaf ears. Two days before the storm, it appeared a military launch would ensue. As fate would have it, nature intervened on our behalf. In just fourteen hours, the electrical power grid throughout China and the eastern two thirds of Russia was destroyed or heavily damaged from electromagnetic activity. Electronically speaking, North Korea was thrown back to the Stone Age. The U.S. on the other hand suffered far less damage, at least comparatively speaking. Steps had been taken, at the urging of Dr. Roth and his colleagues, to insulate the relativity small number of electronic devices still in working. The balance of power had once again shifted; as far as advanced systems were concerned, no entity had a competitive advantage over another. Even if our enemies wanted to attack, they no longer had the capacity or resources needed to carry it out. American ingenuity was put to the test and it proved successful.

Short Term Remedies – Some Possibilities

As the Vice President reported, a large portion of our electrical generation and distribution capabilities had sustained heavy damage; although our biggest hurdle was replacing nearly all of the electrical transformers within and throughout the transmission network. Upon initial inspection it was determined that most of those units weren't repairable. As it turned out that wasn't totally accurate. By sheer luck, and the persistence of several bright technicians, we learned most units were still useable just not at the level they were originally designed for; Phase-One and with some modification, Phase-Two—current—were still a possibility. Larger electrical users, like industrial firms who use Phase-Three were out of luck but individual households and small business might yet get service. We were at war, not with a foreign entity, but in a conflict to regain a way of life. Every human being with a technical background and expertise in this area was put to work. Once mobilized, nearly all transmission and transformer devices were assessed as to their capabilities. Remarkably the process was completed in just ten days; an accomplishment beyond belief. A game plan was put together.

The plan's primary objective, restore electrical service to most of the country, be at a reduced level, in the shortest time possible; but no one, not even the experts, ventured a guess as how long that would take. The skeptics doubted it could be done, the optimists thought six months, maybe a year. Socially and economically the country was falling apart; the technical leviathan we created had fallen down on us and we had to dig ourselves out. The effort was divided into regions; population centers and rural areas. Efforts would be applied equally; at least that was the aim. It seemed nearly all broached the matter with maximum determination. While this was on going, Dr. Roth, his daughter, his son-in-law, and scientists at Stanford and MIT were working on a short term solution for the communication network.

It was a challenging task. Almost from the beginning, they tried in great earnest to forget what they knew about micro and radio waves—

thinking had moved outside the box. They repeatedly asked themselves was there any way at any level micro and radio waves are not affected by electromagnetic impulses, i.e. solar storms. In an attempt to answer that question, they engaged in a series of experiments to test whether it was even a possible. For nearly a month they struggled with that question, without success; they were ready to abandon it entirely. However, Patty, Roth's daughter couldn't let it go even after her husband insisted on it. One night, well past midnight, she awoke with an idea. Not wishing to waste time, she got dressed and drove to Stanford to test the thought. Much her surprise Roth was still there; he'd been there all night; he said he couldn't sleep. So here they were two half asleep scientists, who could hardly keep their eyes open, conducting electromagnetic experiments, using of all things a portable radio and a old broken down microwave oven. Had the two gone mad; to many the answer would be yes, but not them. The conclusion proved easier than even they expected or could have imagined; putting it into practical application on a large scale was way more daunting. Things to, were happening in Washington.

The President's doctors announced he could resume his regular duties. That was good news, except for one important fact; he wasn't up to speed on how things stood with respect to the crisis. Also, there was some question about his mental fortitude that is doing what was necessary to get the country back on its feet. Remember he was the one who dragged his feet when Roth told him about the pending crisis and when he finally decided to do something, it was already too late; the electromagnetic storm hit us before anything got implemented. To his credit though, the President admitted his lack of knowledge and asked the Vice President to assume his duties; he needed to rest, as mentioned before. However before the President resumed his responsibilities, the Vice President had gotten the Congress to appropriate the needed funding. She also assigned technical specialists from each branch of the armed services to aid the scientific community as much as possible. Furthermore she instructed the Secretary of the Treasury and the Chairman of the Federal Reserve to work with the financial sector in developing a highly efficient manual system for handling financial transactions. Since the storm, the system for managing money, deposits, withdrawals, making loans, buying and selling stocks, ordering life necessitates, e.g. food, clothing, medicine and alike, was in shambles. Even paper tracking that some organizations used rapidly got convoluted. It was a huge task even for the best and brightest. Besides the technical obstacles we faced, maintaining social order proved tougher. America was in turmoil; the public in an ugly mood; there was no getting around that fact.

Social Disorder - Here

*I*t wasn't merely a handful of people who had their lives turned upside down, entire communities across the country were in the same predicament; we're talking millions of people. It didn't matter if you barely had enough money to pay the rent or someone who owned a mansion; all were in the same boat. Industries were shuttered; without power and raw materials production was halted, so workers were sent home. Since the financial network was shut down state government wasn't able to distribute unemployment dollars. The Post Office as well wasn't able to process and deliver the mail in a timely manner, if at all. Requests for public service; police, fire, and ambulance were also affected. There was virtually no backup system in place to replace 911 when that and regular phone service went down. In most cases, people would actually drive to government offices and ask for help. Law enforcement had the most daunting task of any government body.

When life gets disrupted and needs aren't met, people often get upset and do things that they might never do under normal circumstances, but these weren't normal times. By comparison, conflict was relatively minor during the day compared to nighttime hours. During the day most altercations amounted to hollering, pushing or shoving but no full scale brawls. However as night fall approached things changed. Perhaps the activities of the day or should I say the lack of them wore on people and eventually needed some kind of release. It was physically impossible for police and sheriff deputies to respond to every conflict. Most fights involved men, but women were also part of the act. As time went on, fights weren't limited to punches and kicks but escalated to using knives, scissors, broken bottles, razor blades, hammers, and metal pipes; and yes firearms. Injuries ranged from minor cuts and bruises, to puncture wounds, and bullet holes. Pleas for calm were mostly ignored. The order of the day was disorder. It didn't take long before jails became overcrowded; rule of law was noticeably absent. Violence, however, wasn't confined to streets and workplaces, it hit home as well.

Spousal arguments became increasingly frequent. Screaming and shouting often evolved into throwing objects, with threats thrown in for good measure. Men weren't the only aggressors women started their share of arguments. Sadly some of those disputes ended tragically, with one or more of the participants severely injured or worse yet killed. Children to, weren't immune, in many ways their actions paralleled adults. Society, for all practical purposes had broken down. Even the religious community struggled to help. A footnote, nearly all our resources, unfortunately, were allocated to solving technical problems, little if any were employed to reduce social unrest. What is left when the former is solved? That was a problem which needed to be addressed and addressed immediately. Given the events unfolding in the world, it became even more critical.

Social Disorder—In Other Countries

*I*n nearly all countries severely impacted by the latest electromagnetic storm social disorder and unrest became the new normal—even more so in China. Robbery, assault, even murder reached epic proportions—people had to protect themselves. All too often, military and local police would leave their post to attend to their family and other love ones. Essentials, from the start, were hoarded in record numbers. People who had resources to buy things were usually charged adsorbent prices. Once bustling economies, ground to a halt. Corruption was everywhere. The politicians were quick to suggest who was to blame.

*M*any blamed the United States and her allies. They claimed the sun wasn't the cause; likely some kind of electronic weapon was cast upon them. What they failed to consider in making their accusations, how the United States and so many countries around the world were similarly affected. Then again, political accusations have little correlation with the truth. In a couple of instances, war was even declared on the United States, as foolish as that sounded. Those countries couldn't mobilize enough resources to meet their own needs much less muster enough to launch an attack. When the posturing finally diminished, the reality of circumstance became all the more real. Whether we or they had any willingness or capacity to help one another was anyone's guess. Despite all the chaos we alone started to make head way.

Incremental Progress

*O*nce was we figured out how we could generate and distribute electrical power, be it at a lower level—absent Phase-Three—service to house and small businesses was restored. Furthermore, power plants once shuttered were brought back in service, providing they could obtain uninterrupted fuel supplies to generate the electricity. Also helping us was the in line phone system; it suffered just minor damage from the storm event. Conventional phone service had virtually disappeared; less than twenty percent of the homes and five percent of the businesses in the country still utilized that service. Virtually every market made the switch to wireless. People still using the old system could communicate with others using the same system. A week after the disaster, a massive effort was made to bring the service up to capacity; when necessary, new cable was laid to critical centers. As additional capacity became available, the once lifeless system—with only twenty percent of utilization—soon reached full capacity. As one might expect, it wasn't unusual for overloads to happen; busy signals became all too common. To reduced the likelihood customers were told to limit their calls to five minutes and place calls only in the evening—after 7 pm; businesses were advised to limit calls to five per hour and none after seven in the evening. Also the number of new customers was severely limited. On the creative side, a paper messaging system was developed; similar to a telegram. Given the circumstances, it worked rather well; you could call it a 'back to the future'. While all of this was happening something rather odd took place.

*I*n a time when half the people grew up with cell phones, emails and text messaging, face to face, not Facebook, Skype or Selfies, made a comeback; mostly out of necessity. Actually talking to someone, face to face, without an electronic go between, seemed natural;

something the old timers knew well. People started to learn more about each another; more than they ever imagined. Folks who had the opportunity to interact found it beneficial. All shared a common problem—electronic interruption—it gave them another avenue to explore new ideas or revisit old ones. Instead of competing with one another, which most did their whole life were now working together for the greater good. If anything could curb hostiles, working together and getting along just might. It seemed natural, you didn't have to think about it; you just did it. It was certainly an idea the politicians needed--working together, not competing for self interests.

As things started to come together in parts of the country, matters in Washington and to a lesser extent in the state capitals, relations were less than cooperative; working together, that was foreign to those people—the elected. Politicians spent more time debating about who gained or lost from a particular decision than making the decision. Leadership was in short supply. Dr. Roth with help from the Vice-President, had the best chance of getting them going in the right direction; Roth wrote the remarks she would deliver to the Congress.

Never in the history of the United States had a sitting Vice President delivered a message of this magnitude; and certainly not before a joint session of Congress. Roth, as a guest of the Vice President, was seated in the front row directly in front of the Speaker's platform. The VP's remarks were schedule to last thirty minutes, including a short summary on how things stand. She began with an assessment of the challenges now facing the nation. Instead of appearing concerned with our situation, most people in the audience seemed indifferent even bored by what she was saying; even when she pointed out what we had to do to get back on our feet, many appeared ambivalent. For whatever reason her message wasn't getting through and she knew it. Out of desperation, she called on Roth to come up and explain things more completely. Her request caught him off guard; he wasn't prepared to speak, and had no text. As such he took a different approach; he observed the mood; Congress needed a jacking up.

He began his remarks by saying he wasn't invited here to give a speech, yet here he was standing before the nation's leaders, all five hundred and thirty-five of them. It would be nearly impossible to paraphrase his comments; instead here is a select portion. "Ladies and gentlemen, madam Vice President, I stand before you as a scientist, as a father, as a veteran, and most of all as a citizen who loves his country. I like you have reaped the benefits and rewards that progress through technology has given us. But nature, not human beings has taken some of those benefits away. It has disrupted our lives and now threatens to turn us against each other.

Prior to May 1, I warned the President and several of you in Congress such an event was probably, and if and when it occurred it would have a major impact; I also warned a subsequent storm would likely follow, which it did, bringing even greater destruction. I presented my warnings not as a way of getting recognition or additional funding for research in my noted profession. I gave those warnings because of my deep admiration for my country and my fellow citizens. What I observed earlier, as the Vice President spoke, that a great number of you didn't seem to share her conviction, and that concerns me.   Pretending something doesn't exist, simply by ignoring it, denies reality. Things never fix themselves. Every day I hear stories about people exploring new ideas to make things better, but ladies and gentlemen they can't do it alone, they need your help, we need all hands on deck. I don't give a damm if some special interest group or some lobbyist doesn't like what you have to do or whether or not such action might be used against you in the next election. It's time to do the job you were elected to do in the first place; any fool can argue and complain, cooperation takes work and if we're going to get through this we have to work together. So let's get going, now that's all I have to say, good day." As Roth left the podium he didn't know what to think, he knew what had to be said was said; the hell with the consequences. As he was about to sit down, the House chamber promptly erupts in thunderous applause; everyone was on their feet. It was the last thing he expected. Even the Vice President came over to personally thank him followed by an array of Congressional leaders. Later the President even invited him to the White House for supper. The VP was also invited. Later that night, following a nightcap, he began having second thoughts. He never wanted the spotlight; he knew it was a dangerous place; besides he had more important things to do.

*I*t didn't take the media long to react to Roth's comments. Even the word hero was brandished about; spun as a story, about a regular guy who stood up to the "bad boys". Roth wasn't prepared to deal with all the adulation; he was never comfortable in the limelight. At the very least, it gave the public something to focus on; maybe things were turning around after all. Then a week passed, still no action; what happened next caught many in Congress by surprise. People, by the thousands, millions would be more accurate, took to the streets; demonstrations were taking place everywhere, in every part of the country. Protestors made it abundantly clear, they weren't going to sit idly by while government officials sat on their collective butts and did nothing. Attempts to appease the public were unsuccessful. Moreover, officials were hounded wherever they went, in stores, at the airport, on the street, at church; and it was having an effect. Those up for reelection were hit especially hard, hearing phrases like "start packing your bags you'll be soon heading home" or "and a nobody just like us." Since instant communication no longer possible, to the degree it was before, officials now couldn't speak to the electorate directly, in mass. Roth still managed to enjoy a brief respite from all the commotion; then demands rolled in.

Three days after delivering his "lecture" he was back at Stanford lab doing experiments with Patty and his son-in-law's help. He relished in solving problems, and the country had a host of them. In his absence, Patty learned a few things about AM waves. Most importantly, if the frequency is reduced by just two percent within one quarter a quarter of a second interval, electromagnetic assaults had no effect on said waves. That revelation could a key to restoring, at least partially, our communication system, at least in the interim. However, they still had the task of overcoming signal disturbance, a.k.a. static. Roth had several ideas how to deal with it but it would take more time to accomplish it. Somehow news of their breakthrough got out. Almost immediately letters poured in, and even a call; somehow a political activist discovered the number for the Stanford lab—the line connected directly to Roth; it was put in years before in case of a natural disaster; more puzzling, just a few people knew of its existence; and they were high level officials.

Scientist to Reluctant Spokesman

At first, Roth refused to talk or say anything; he simply told the caller he had more important things to do; he was scientist not at a political jabber box. For a few days, that strategy worked, then friends and colleagues called; they assured him engagements would last just a few days; he'd be back in the lab before he knew it. After considerable arm twisting, he reluctantly agreed to speak at five venues, not as a lecturer per se but rather as a resource person. In any case politics would not be the focus. Oddly enough, all those engagements would be held at Stanford; that put his mind at ease.

The theater where he would speak seats 3,500; in his mind he thought two maybe three hundred people would show up for any one session. For his first engagement he chose to enter the theater via the underground tunnel which was connected to the science building where he and Patty did their experiments. Just twenty minutes before the presentation was to begin, he reluctantly hangs up his lab coat, straightens his tie and heads for the tunnel. An aid catches up with him along the way and mentions attendance is larger than expected. When he arrived at the theater, he waits patiently backstage until the University President introduces him. It was a glowing introduction. When Roth walked to the podium he was surprised by the size of the audience; every chair on the floor was filled, in the balcony and they were only a few seats left. Everyone was standing; the applause, deafening.

Once the crowd quieted, he interjected a touch of humor, which was for him was unusual. He said with a smile, "I must apologize I seemed to be in the wrong room." That brought another round of applause. Now he was nervous as hell, speaking to Congress was one thing, talking to the public that was another matter entirely. He reminds the audience he isn't there to lecture rather reiterate where we find ourselves, the progress we've made so far, layout the challenges that lay ahead, and most importantly give you an opportunity to ask questions. He wasn't there to berate anyone's opinion or minimize the affect the storm put on personal or professional lives.

Roth had no intention of sugar coating anything. He said nature, by way of the sun, had crippled our communication system, not only here in the United States but in other parts of the world as well. Our heavy reliance on a high tech, state of the art, got to have right away communication has come back to haunt us. The threat of an electromagnetic storm has always been with us, but routinely we discount the probably of its occurrence. Necessary precautions could have been taken, but because of the time and money got in the way we chose not to. The game of should've, could've, would've serves no real purpose except to let it occur again. He emphasized, there is just one way to go, that's forward, straight ahead. The solution lies within us. Progress has been slow, but we are getting somewhere. He mentions the partial restoration of the power grid, revival of conventional phone service, although quite limited and promising research on alternative communication methods. However what bothers him the most are questions from the audience; in the classroom he's in charge, but here not so, and that frightens him. In the allotted time--an hour and forty-five minutes--he responds to twenty nine questions.

Space doesn't permit mentioning all, what follows are the most pertinent, in random order.

1. When will cell phone and internet service be fully restored and will it work the same as before? *Answer,* in the next three or four months, the answer is no, will it work the same, probably not, a host of systems that go into providing those services were either destroyed or badly damaged, the amount of equipment that needs replacing is enormous and will cost a great deal of money.

2. Why do we have limited electrical service, isn't electricity uniform? *Answer,* the storm destroyed much of the electrical distribution grid and the transformer transmission system. Fortunately we're able to modify the system to provide adequate supply to meet Phase-One and two needs; our capacity to generate

Phase-Three currently is quite limited so we have limit to access to the most critical areas; hospitals, fire, police, ambulance stations and municipal utilities.

3. Why can't we access our financial records, transfer money or otherwise conduct normal banking activities, along with buying and selling stock? *Answer*, most of those processes require cell, internet, and electrical service, i.e., Phase-Three power, services that are not available or very limited. Given our situation, security is not guaranteed.

4. Why are food, gasoline, and other essentials in such short supply? *Answer*, it goes back to what I said earlier, distribution of goods and services are highly dependent on technical systems for ordering, producing, shipping, and yes financing them; remove or otherwise

eliminate any one of those three the whole system slows dramatically or stops.

5. When you were invited to a joint meeting of Congress with the Vice President, were you asked to speak at the time? *Answer*, no not at all, I was there to lend support; the Vice President was to do the talking.

6. If you weren't prepared, and I assume you weren't, why did you agree to speak when asked, and I must say it was an impressive presentation. *Answer*, as you probably know, the Vice President struggled to get her point across with respect to what the White House and Congress had to do, as such she invited me to the podium after giving me a look of desperation. I felt it was my duty as a scientist, as father, as a citizen to make our point in terms they could understand.

7. Did you have any misgivings about the manner in which you spoke to the Congress? *Answer,* no not at the time I was there, I simply said what needed to be said, pure and simple; only later did I question whether I came on too strong. We need to get things done to restore our way of life; we needed Congress to pass legislation to achieve some of those goals and up now they aren't doing a damm thing and that upset me.

8. I was told you're conducting experiments at the Stanford lab which involve the effect electromagnetic activity has on radio waves, could you tell us what that consists of and what you hope to learn from that? *Answer* by studying how electromagnetic waves or pulses alter radio signals—e.g. cell and wireless internet transmissions, we hope to learn at what level or range they do not. Right

now we're studying AM (audio modulated) radio waves and given what we learned so far we think we can alter the frequency and timing interval and bypass nearly all electromagnetic interference. If we can effectively control the static inherent in AM, a short term communication alternative is a possibility.

9. When do you think such an alternative could be ready? *Answer*, I wish I could give you a definite answer, the truth is I don't know. My daughter, my son-in-law, and I are committed to working on this full time. We are not out to play games, my family is suffering as much as yours.

10. Dr. Roth, do you plan to run for elected office at some point? *Answer*, absolutely not, I'm a scientist and a teacher, my place is in the classroom not the stump.

*R*oth had no idea his comments or for that matter his responses would make any difference. The truth was they did, because the very next day several bills were introduced in the House of Representatives; officials even agreed to bypass Committee review and send them directly to the floor for a vote. He respectfully declined additional speaking requests once the last one was completed; he needed to get back to the lab. Thankful those making requests understood the need for added research. His return lasted all of four days; then the firework started.

*W*hen the proposed bills were read on the House floor, the financial and communication lobbyists protested loudly. The first bill directed all communication system customers, e.g. computer, television, internet, etc. not be charge for any service not afforded them, regardless of the reason, starting with the first day that service or services is interrupted. In addition, said companies must guarantee to customers that once service is restored, it shall be at the same price charged preceding said disruption. Furthermore the access charge or service fee imposed during the interruption shall be capped at $5.00 for all customers. Those rules, with few exceptions, were the same for telephone providers, cell and land line, accept that any customer charged an access to service fee may cancel their association without penalty, i.e., able to purchase that same service at some future point if they so chose. Finally, all phone companies currently offering land line phone service shall, at the maximum extent possible, increase technical capacity.

*T*he second bill was directed at the financial community, therein severely limiting activities of specific institutions. Effective immediately, all stock transactions, whether by purchase or sale are hereby limited to three per twenty-four hours, and cap each transaction at one million dollars for financial institutions and a hundred thousand dollars for private individuals. Banks, hereafter, may not penalized a customer for withdrawing funds from their institution, provided said withdraw does not exceed fifty percent of that customer's holdings presently on deposit. In addition, they are not permitted to adopt new fees or charges for service, nor recall outstanding loans, provided all previous payment obligations have been satisfactorily met. Finally all banking and all other financial firms shall develop and immediately put in place an efficient and effective paper and auditing system to track all transactions formerly managed electronically, until such time as the electronic system is fully restored and 100% operational.

It was a well known fact, each sector enjoyed a free ride for years; Washington had simply rubber stamped everything they asked for, now they were being told what they had to do and they didn't like it; spoiled kids being told to clean up their room and do their homework.

They blamed Roth for their predicament; after all he was the one who pushed Washington to act. The truth, Roth facilitated the necessary push; most politicians had already been accosted by constituents back home. The public had become livid and put the fear of God in them, even threatening to boot them out of office if they didn't get to work. Much as he hated the idea of getting back into the fracas, Roth needed to set the record straight and restore his reputation. He took to the radio a few days later; there was still one network capable of broadcasting coast to coast.

"Yesterday, members of the communication and financial community publicly stated I was the reason restrictive measures were put in place by the Congress. That accusation is incorrect; I intend to set the record straight, I acknowledge suggesting Congress do something to get the country out of its current dilemma, while assuring every American is protected from unscrupulous conduct. I had no part nor gave input in the drafting of that legislation now before Congress. If those bills give companies pause, I suggest they're more interested in their own interests than those of their customers; my allegiance lies with the later. Presently I spend by day doing research, with the hope of finding a solution to our problems. Now that's all I have to say, thank you for your attention, good day." Response to his remarks was immediate and overwhelming; support came from ranking members in Congress and the White House. That said all Roth wanted was to be left alone. However, there was a demand he couldn't avoid.

MIT wanted him back; days earlier they were awarded a multi-million dollar grant for solar research, including related experiments. NASA was also a participant. With some reluctance he agreed, with one stipulation and if they declined he would turn down the offer. His daughter and son-in-law would have to accompany him, assuming Stanford granted them a leave of absence. Thankful they did, as long as university receive appropriate recognition. The focus then shifted to the political campaigns.

Political Campaigns – Flash from the Past

*B*efore May 1, political campaigning enveloped the country in the usually manner, has it had for past twenty years. There were political ads on TV, on radio, and the worldwide web bombarding the electorate. Though news coverage among media giants had gotten more creative, or should I say more fabricated, the relative truth of things a secondary concern. Talking bubble heads "experts if you will" were everywhere, vainly trying to put a spin on things. However this year the political race for President was way more competitive than previous ones. The presumptive candidates from each major party had virtually identical poll numbers. The Republicans held their convention the second week in July; the Democrats a week later. Just minor disruptions occurred following the May storm; as such they had little or no impact on the race. The Republicans nominated the President for a second term; the Democrats selected a relative new comer, by the name of Mike C. Nelsen. A former two term Congressman and present Governor of Minnesota. Most political pundits considered him extremely intelligent, with a charming personality, and a well regarded as an excellent administrator. Whereas the President was considered by many as a "steady eddy"; the conservatives loved him. He projected an image that of a stately professor from a private school back east. Once the devastating storm struck America and Europe, the political landscape changed dramatically, virtually overnight. America was posed for change.

Political ads, except for a handful of local and regional newspapers, had virtually disappeared. Television and nearly all radio stations had been knocked off the air—hence no means to get a message out. Wireless and cell phone service were out as well. Campaigns that raised millions dollars for advertising had no place to spend it. Political campaigning had reverted back to the 19th century. In the age of instant everything which people had grown accustomed to, the political machine could not accommodate. The rules or circumstances of the game had shifted. A return to yesteryear, old time campaigning became the new reality; the age of 'stump speeches' had returned, literally speaking. In earlier times, candidates for office would venture from town to town to deliver a message. To be seen and heard, politicians would often stand on a tree stump to deliver their speech. Needless it say traveling from town to town was a slow and grinding process, requiring plenty of stamina. As candidates moved from state to state, political flyers would always precede them. Without question, 'stump speaking' greatly favored Governor Nelsen. He was attractive, physically fit and was good at delivering speeches. More importantly he understood the situation nearly all Americans found themselves in. The President on the other hand, came across as stuffy and seemingly confused to how to improve things. Nelsen's words sounded more like Roth's message to Congress.

On campus at the University of Wisconsin—LaCrosse, his words were emblematic. "Today we find ourselves in a situation not of our own choosing we are at the mercy of nature, a power far greater than anything man could create, with the exception of the atomic bomb. Whether we could've prevented all this from happening is best left to another time; it serves no purpose to play the blame game; we need to put our energy into getting back to where we were. We need to listen to one another; now is the time to do just that; it shall make us a better people….If you elect me President I'll be beside you every step of the way; we will struggle together. As others did

before, we can persevere and we will. Divided
we accomplish little; together we can accomplish
things. Thank you for your attention, good day."

A Change in Leadership – Success and Challenges That Lay Ahead

The 2020 election would be the first election in a
hundred and fifty years that big money played
no part in the outcome. Big money needs mass
media to spread its message, that is who is or is
not the best choice; people with money just had
no way of spending it to make a difference.
Conventional wisdom suggests it benefits
incumbents; but it didn't; but not this time. In
the House of Representatives, just thirty-seven
House members got reelected; in the Senate out
of the thirty four members standing for
reelection just six were picked to return. On the
state and local level, the outcome quite the
opposite, nearly ninety percent of incumbents
held onto their job. In the Presidential race, the

President lost big time; the margin 71 to 25. Governor Nelsen was headed for Washington. Vice President Webber, in an unusual move, would stay there after the inauguration, at the request of the President elect. She and Roth were asked to join his team as technical advisors. Nelsen was impressed with their work through much of the crisis.

Now he had to let the America public know the extent of our problems. An exhaustive report on what was and what wasn't working, our progress in restoring certain elements and a realistic time line to make future improvements, absent political spin; the public needed to know the unvarnished truth.

On January 27, 2021, one week following inauguration, President Nelsen along with Dr. Roth and former Vice President Webber collectively held a news conference to discuss the present state of our communications network and its impact on everyday living; without a doubt it was a sobering commentary. Following his opening remarks the President turns the podium over to Roth. What follows is a summary of his remarks.

Technical Facts:

1. As of yesterday, 16 out of a total of 147 satellites owned and operated by U.S. entities are operating, some only marginally. Of that amount, just 7 are fully operational and thus dedicated exclusively to national security and matters related to health and safety. Another 31 satellites are in various stages of development but no specific timetable has been prepared for their deployment, our hope is to have three possibly four units operational within ninety days, but that is not etched in stone, these are highly complex units and we have to insure that all the

systems in place to prevent future disruptions if another storm occurs.

2. As of this moment, we have no exact cost estimate to repair or replace damaged units—satellites.

3. With few exceptions, nearly all cell phone networks are down and unusable. We simply don't have enough satellite capacity to handle demand with any degree of reliability. Reliable service will not be available until we have the technical capacity to handle it. We hope the land line system will meet some of our needs. Prior to the storm, that system was operating well below capacity, a mere eighteen percent. The reason, the businesses and individuals had switched to cell or wireless service. As a result the system was never properly maintained. Then all of sudden demand explodes; as of today the

system it at capacity. We're hoping we can increase it further but for now it can't handle any more customers. Moreover, customers may have to wait for a dial tone on occasion.

4. The nation's wireless network, with several exceptions, is down and unavailable, principally because of our satellite situation. Television, radio, GPS operations are affected as well. However, there are some reports that a handful of systems somehow managed to operate but reliability is much in question.

5. The nation's electrical generation and distribution systems proportionately are in far better shape but to a point. Right now we're able to provide Phase-One and Phase-Two power; Phase-Three capacity now stands at five percent, as such it's dedicated to essential systems when possible. Fully

restoring Phase-Three will take longer; replacement transformers aren't readily available; we need hundreds of them.

6. The country's financial and economic systems to have been impacted as well severely I might add. As a nation, we have become too dependent on technology to meet everyday needs. Instant communications, whether for voice, entertainment, messaging, money exchange, and others allowed us to abandon old management systems; I'm not saying that was a bad thing, what I am saying is, if that system goes down or is unreliable there is nothing readily available to fill the void, at least to the level were accustom, which leads me to something else, personal circumstances.

Personal Impacts:

Without question, daily life has changed; the speed at which we conduct our affairs has been altered, because the system we developed and used to meet expected demands has failed us. Everyone, the rich, the poor, individuals, families, and businesses all find themselves in the same circumstance. The fact that I'm a scientist doesn't protect me from those impacts or our leaders. All here on the dais feel your pain.

When you get up in the morning, you wonder if the lights will come on, wonder whether water will come out of the faucet or whether the furnace will turn on. The most basic of concerns are now paramount. For the worker, you worry about your job and whether it will be there tomorrow, next week, or next month. Phone disruptions prevent you from talking with family and friends absent face to face contact. Entertainment is reduced to the most basic of elements, like playing games on a field, reading, or failing those, worrying. Then there are money worries, like accessing bank accounts, stock portfolios, or not having a steady paycheck. Most importantly you worry about meeting daily needs, such as having enough food to eat, gas for the car, medication to maintain health or for personal safety; I can assure you, everyone up here is aware of those things.

The challenges we face requires us to reevaluate things we once took for granted. To get through this, we must concentrate on what's available. On that basis, we must use those resources we have available to find solutions. In a moment, the President will tell you what government has done or will do in the days ahead, President Nelsen."

"Thank you Dr. Roth. Since the election, I've spent a great deal of time studying the crisis; specifically those actions the federal government has undertaken, and I might add they have been extensive.

1. Congress has allocated funds to satellite manufacturers to enable them to step up production on replacement satellites; those companies and their subcontractors are working 24/7. Prior to the electromagnetic storms that devastated their equipment, only four replacements were in our inventory, with three more in various stages of production. The first of those replacements were launched into space on December 30; subsequently each of the remaining units will be launched thereafter in monthly intervals. Right now, we have just one launch vehicle in service; it will take another six months before another is ready. I can assure you we'll do everything possible to expedite things.

2. As Dr. Roth mentioned, our electrical generation and distribution systems were heavily damaged. Through

tremendous efforts, people working in that industry have managed to restore low and intermediate service in record time. Restoring Phase-Three power is going to take far longer; we simply don't have the inventory to replaced all the damaged units, even at full production, it takes nearly a month to produce and install one. Customers with the highest priority, e.g. defense installations, hospitals, emergency services, public utilities, etc. will have capacity as it becomes available. I wish we could bring everyone on line at the same time, but we simply can't.

3. The previous Congress along with my predecessor was criticized harshly for implementing emergency and highly restrictive measures upon industry. Their rational for doing this, protect the financial resources of the average citizen, from those who would take

advantage of the situation to extort hard earned dollars. There was simply no other way of protecting those funds in another way. I know those rules pose an inconvenience, but they're necessary. The rules will be rescinded, once we're confident your money is adequately protected. In the meantime, with have installed dedicated phone land lines in three bank locations in communities whose population is 5,000 or more, to answer your questions. Treasury Department personnel will be standing by to receive your calls, from 8 AM to 8 PM Eastern Time, Monday thru Saturday.

4. Public safety will be a top priority for this administration, as it was for the previous one. The National Guard is prepared to quell any violent disturbance in conjunction with appropriate police and sheriff

officials. If conditions warrant, the regular Army is prepared to assist. Keeping violence in check and protecting property is paramount. However, public assembly will not be curtailed unless it turns violent; a determination made case by case.

5. As to my last point, I like to compliment Dr. Roth and his esteemed colleagues at NASA, MIT and Stanford for their hard work. Tomorrow I'm going to issue an Executive Order therein directing that NASA be given the necessary resources to regularly track and monitor solar activity of the sun and other elements in our solar system and to report those findings each month to this office and the Congress, those finding shall include specific courses of action necessary to minimize the effects that may result from future

storm events. In addition, the federal government will provide the necessary funds for research and development of new or otherwise modified methods for transmitting electronic signals to consumers and companies, via transmission stations, intermediate units such as cell towers, satellites, or other devices, safely efficiently.

One final point before I will open it up to questions. In the days and months ahead we face huge challenges. I would love to say, everything will be back to normal shortly, but I can't. The normal, we had, has disappeared. You and I will need to define the new normal, and by maximizing opportunities we'll come to recognize and also take advantage of them. At this moment, opportunities seem obscure, but they are there out there; it will take work to find them out. My family, like yours, experienced much inconvenience: job loss, ready contact with family and friends, and loss of services to name a few. To suggest government doesn't care or isn't doing anything is just wrong. We fell together, we'll rise together; now your questions."

For the next two and a half hours, the President, Dr. Roth, and former Vice President Webber fielded a host of pointed questions. Repeated attempts to champion the blame game were dispelled by the President. Overall it was exhausting. That evening the President, Dr. Roth, Mrs. Webber and a few staff members dined together and critique things. Collectively, they agreed the biggest challenge was not technical—that's a given—but rather keeping the public in a positive mood. A mass revolution would certainly tear the country apart.

Finally Some Good News

Despite all the bad press, there was some promising news to. Initial assessments of the electrical distribution system concluded none of the damaged equipment was repairable; that was not totally true. Fortunately several very bright technicians determined a portable device could bypass damaged components, and although such portables were in short supply, although new ones could be produced rapidly. With some luck, about twenty percent of the country could have Phase-Three power in as little as five weeks; it would take several more months before additional service could be afforded, although some isolated areas would be excluded and there would be additional challenges. It was decided no announcement would be forth coming, thus avoid undo expectations in the event the method doesn't pan out.

News involving the damaged satellites presently in space also brought a glimmer of hope, but it was far too early to report a break through. Initially scientists and technicians thought the multiple storm events had destroyed most if not all the circuitry onboard, but that wasn't entirely true. Power to operate the satellites came from the sun, using an elaborate solar grid, in layman's terms a high tech antenna. Scientists concluded they were not working. However, every satellite had a smaller backup grid onboard just positioned in a different direction. In the event the primary unit wasn't operating, the backup or secondary system was designed to come on line to insure the craft maintained its proper orbit. What the scientists didn't know, whether that device was designed to send, receive and process signals received from Earth. Since that device only came online when an interrupt signal was received from the main grid; which obviously it didn't, could that device alternatively receive the same signal from Earth, hence restoring electrical power. It was something they never anticipated; so no procedure was ever developed. It took engineers nearly a week to come up with one; it was tricky. The signal would have to be sent to each satellite individually; all had their own encrypted security code. Complicating matters, in order to receive the signal, the satellite had to be at a precise location in orbit. Communication with the first three, were unsuccessful; the fourth produced a weak response; it was assumed it like the others severely damaged. The news was disheartening and the procedure nearly scrubbed. However one person was allowed to continue testing. Two more failures followed; the seventh successful. Within an hour, that unit's secondary grid operates at maximum capacity.

Although satellite seven had electrical power, engineers didn't know how effectively it could send and receive signals or if it was capable of maintaining consistency. They also worried about the longevity of that secondary grid; it wasn't design to operate at that level for an extended periods. The last thing they needed was to over task the unit and have it fail. It was a short term fix, nothing more. In total, thirty-nine satellites were brought back to life, though operating on life support. Given the complexity involved, a gag order was put in place, there would be no public announcement, only the President, Dr. Roth and several colleagues at NASA and MIT were informed.

While engineers determined the quality and reliability of each satellite's signal, Dr. Roth and his colleagues debated what role those satellites would play. Security, public health and safety topped the list—reliable units were earmarked for that purpose, any more would be held in reserve. Debate about what to do later proved contentious. Some wanted additional units earmarked for general cell phone use; the problem with that, deciding who would and who would not get service. In the end it was decided, the banking industry, under the direction, supervision, and control of the Federal Reserve would determine their use, if and when they became available. However, it was also agreed, should industry abuse the privilege, the federal government could revoke that privilege.

*R*oth and his colleagues presented their recommendations to the President for his review. He concurred with their conclusions, with one exception—that involved the legal community. He felt lawyers may to try to block said action, claiming their respective clients—the satellite owners and lessees—had legal right to use their satellites as they saw fit. At the moment, he wasn't sure the Emergency Powers Act gave him the authority to issue such a directive. Before signing the order, he consulted with the Attorney General. In another twist, the relative calm the country was experiencing the past three weeks was about to change and change dramatically.

Discontent Turns Violent

*T*he plight of many had reached a breaking point, with little or no money to buy basic necessities, such as food, clothing, or medicine left them frustrated and angry. With anger, violence typically ensues, which it did. At first it was confined to larger cities but rapidly spread to smaller communities, then to rural areas. Public demonstrations became shouting matches, primarily about who is responsible and what should be done to fix things. When shouting didn't work, throwing objects became common place, followed by outright brawling; injuries ranged from mere bumps and bruises to broken bones. Looting also became part of the commotion. In some cases, buildings were set on fire. At times, law enforcement was able to contain things, but eventually their efforts failed. Police and fire personnel were in the same boat as were their friends and neighbors, only less so. That apathy may have kept them from doing a better job. Only the National Guard and select companies of the armed forces—Army and Marines—proved more effective, yet the message was clear. The nation was coming apart, a piece at a time. We expected instant everything, while possessing a lack of patience if and when something wasn't working right. This time there were no quick fix; in our rush to design instant everything, we paid little mind to that fact it might all fail one day. Na Sayers, what the hell do they know. Maybe we thought we were infallible. Somebody has to take the blame for all of this, it certainly wasn't regular folk—the captains of industry and government, they're the culprits; or so the thinking went. Nelsen decided he had to do something fast, but he needed the nation's governors and mayors to help him.

Thankfully after 911, Homeland Security at the request of the President and the concurrence of the Congress directed that a dedicated video and voice lines be installed underground, therein connecting the White House with every State Capital in the nation. Nelsen was about to utilize that resource. Word was sent out to every governor therein asking them to invite all mayors in their state to a joint conference with the President using the dedicated link. The meeting was set for early April; everyone was expected to attend. It took the President nearly a week to draft his remarks including his plan of action. Some referred to as a peacetime "War Plan", we were at war with one another; the outcome uncertain. At precisely 10 AM, on April 5th, the President called the Conference to order.

"Ladies and gentlemen I called you here today because we face a crisis of epic proportion, the largest in history. Through an act of nature, we find our electrical and communication systems completely and totally compromised. Things we were accustom to, are no longer available, at least in the short term, possibly much longer; nevertheless we must plan for the future and think long term. I can assure you, we've put every available technical and financial resource into it. No one can give you an exact timeline for correcting these problems. The truth is we don't have a backup system or the needed equipment to meet even a portion of those needs; having said that, though our biggest challenge is not technical, it involves human nature. As all of you can attest, the public is taking their frustrations out on each other. Believe me when I say, nothing is solved by this type of conduct; i.e. burning buildings, damaging property or hurting one another. Any relieve they may realize is short lived, the problems still remain. In a few minutes staff will present our plan for meeting the 'human challenge'; and you folks will be the ones to take the message back to your constituents. No one human created out plight, however it is our duty as civic leaders to resolve it as best we can and to the best of our ability; as elected officials we have a duty to uphold."

For the next ninety minutes the President and his advisors, including Roth and former Vice President Webber outlined a very detailed plan, inclusive of timelines. Without going into specifics, topics discussed included: financial transactions, such as how individuals and companies can do business with the financial community and do it in a safe and secure manner; coordinate the use of voice, text, and email communication, wherein specific timeslots would be proved, through reservation, to access the nation's highly limited but very operational communication system of the military; of all the challenges currently facing us, and it is the biggest; maintaining the national economy, i.e. the production of goods and services to sustain the workforce; mass unemployment could cripple the economy and produce a major depression, which we might not recover from. Other matters dealt with transportation, public safety, public and private utilities, and finally how to maintain a positive attitude as we work through these matters. Of all the issues confronting the President that one was the most troubling. And trouble was brewing, only hours after the conference concluded.

Discontent Goes Nationwide

*L*ate that afternoon, several militants groups from across the country chose to hold mass demonstrations in four cities: New York, Chicago, Los Angeles, and Washington D.C. on the steps of the Capitol no less. What started peacefully, with speakers expressing concern and a need to work together, soon atrophied into name calling and destroying property as their way of showing their discontent in what was not happening. Innocent people, men, women, and children were caught in the crossfire. Sadly, police and fire officials did little to quell the disturbance; likely empathizing with the demonstrators. As darkness approached, streets became a war zone. Windows everywhere were cracked or broken, you would think a tornado had come through—a human tornado no less. Broken bottles and mounds of paper littered streets and sidewalks, not just in the downtown but also in otherwise quiet neighborhoods; the human toll was even greater. In Los Angeles alone three hundred people were injured in the ruckus; seven were bludgeoned to death. In New York and Chicago numbers were slightly smaller, two hundred and ten and one seventy-six, respectfully, thankfully no deaths. In Washington the total was less than fifty, but the damage to property far greater than anywhere else. The situation was horrific, even if the President wished to speak to the nation—and he did—he couldn't, reliable coast to coast communication was not unavailable. Instead a message was sent via newspapers.

A Worldwide Dilemma —Friend and Foe

Nearly every country had communication difficulties, not only getting what limited infrastructure they had up and running, but insuring its reliability. Transmissions under five minutes were highly reliable; anything longer, no better than 50/50. Even if our enemies wanted to take advantage of us, given the circumstances, they had neither the tools or the devices necessary to carry out such an attack and what if they did, what would they achieve as a result; our situation was no different, perhaps worse. Still that didn't stop them from spreading vile propaganda—principally accusing us of causing all the problems. For now, international relations as well as diplomacy were put on hold. You can't fix a neighbor's fence if you don't have tools.

Demonstrations in some parts of the world were peaceful and relatively calm while ours was yet to reach its peak. By some estimates, and depending on who you believe, demonstrations would run their course in a week, maybe two. A month after the first outburst it finally quieted down—a pleasant calm before the clouds returned, which they did. As luck would have it, if you can call it luck, the storm just moved on to more fertile pastures, to other parts of the world.

But this wave proved much greater than anyone could possibly image.

As it occurred, it grew in intensity. Injuries in the thousands, deaths in the hundreds, and in a few cases in South America it reached into the thousands. Societies seemed to be falling apart; and helpless to stop it.

Just when it seemed there was no more good news, we got some, be it a sound bite, but at this point we were grateful for anything. We were starving; even a parcel was fantastic. Roth, his colleagues from MIT and Stanford, his daughter and son-in-law, would make an announcement. It was a breakthrough everyone was hoping for.

A Glimmer of Hope — be it a Small One

It was long speculated but never proven or for that matter never fully investigated that AM and FM frequency wave lengths could be combined to serve a useful purpose. Then one day, probably out of frustration, Patty overlapped the two outside of their normal operating range, while at the same time bombarding the blend with electromagnet shock waves. Instead of dispersing the signal and rendering it useless, the combined signal continued unabated and seemingly gained strength. At first she discounted the results, thinking it was simply a fluke. In the outside chance it wasn't, she repeated the experiment three times; the results were the same, then she called her dad in. He was ecstatic. Almost immediately, he and the others began talking about to best utilize this revelation. If, and it was a big if, the hybrid signal could effectively transmit to orbiting satellites, perhaps those devices—satellites-- could be reprogrammed and possibly become operational. The only fly in ointment, the batteries onboard each satellite, if the solar recharging packs are damaged or destroyed, the batteries are useless; bottom line, the satellites

162

were space junk. So far, they hadn't tested those batteries, once they discovered standby antennas could be used to get the satellites back in service. Battery power was critical to reprogramming.

For two and a half weeks Roth and others worked tirelessly to manage the newly combined signal. Although strong, it becomes unstable on occasion and they had no idea why. If they couldn't control them, their discovery was for naught. What they learned, simply by trial and error, the speed of transmission is directly correlated to stability of the signal; the slower the signal, stability increases dramatically, the opposite is also true. What Roth and his colleagues needed to know, how to achieve maximum optimization. In the interests of time, they chose to experiment with one of older satellites, that unit was responding to a certain degree but far short of optimal.

*F*or a while, they weren't getting anywhere; too strong of a signal the satellite failed to respond; a weaker signal the satellite didn't perform in the way it was supposed to. They were close to abandoning the idea; Roth was the exception. He personally took on the challenge and nearly worked himself to death, but he found the answer. It was highly technical and defied a layman's explanation. It just worked, that's what counted. Technicians, at Roth's direction, reprogrammed the satellite—it took four days; one down, more to go. What scientists failed to consider in their rush, reprogramming could alter their ability to hibernate satellites should an electromagnet storm reappear. So as not to raise concern, they kept that fact to themselves. Also there would be no public announcement about reprogramming procedures, only the President, several key advisors and scientists directly involved would be informed; they didn't want raise hopes should efforts fail. Only after a thorough and comprehensive testing was finished would it be announced. The electrical distribution system was still a troubling concern.

Numerous pieces of equipment made up the system, much of it had been rendered useless, beyond repair. Production capacity simply wasn't large enough to accommodate need, which was extremely high. Nearly all of equipment in the system was designed to last twenty years; replacement units had to be ordered a year in advance; actually deliver would happen eleven months later; installation and testing another thirty days. It would take at least two, maybe three years to fill all the orders, assuming manufacturers worked 24/7, seven days a week. We had to live with that reality, but it didn't stop crackpots from offering their own solutions, none of which were credible. While scientists were working to solve the technical issues, Congress and the President struggled with how to pay for all of it.

A Matter of Money

*G*iven the large turnover in Congress resulting from the last election, new members had no idea much less an understanding how to resolve the money question. Right or wrong the public assumed the government would pick up most of the tab; newly elected people tried to dispel that notion. It was okay, as long as someone else paid for it. The trouble was, we were fresh out of someone else(s). Nelsen alone had to decide who should pay and how much. After talking it over with his advisors, he decides the communication companies needed to cover the cost of repairing their respective satellites and/or building and launching replacements besides building and installing an alternative communication system. To soften the impact, the President directs the Congress to pass a onetime tax credit for every company affected, at an amount not to exceed forty percent of said cost. He knew he would be criticized, yet he wasn't about to bankrupt the government by paying for everything. The issue of public utilities was another matter.

Major changes were needed, specifically how to generate and distribute electricity. In just twenty years, we went from producing and distributing electricity from a defined area; often less than a two hundred mile radius to much larger quadrants geographically. Today, distribution extends out two or three states and in some case well beyond that. The push to switch to solar and wind generation, popular at the beginning of the 21st century, all but abandoned. Inadequate storage capacity—batteries--was main reason. Power generated from either of those sources had to be consumed almost immediately; absent adequate storage, industry growth became almost impossible. Nevertheless solar and wind was still viable as long as the cost of fossil fuel was high. However at middle of this decade i.e. 2015, the price of petroleum fell dramatically; and therein those fuels resumed their position of dominance in generating electricity. Since most wind and solar infrastructure was in place, the President decided to order an immediate assessment of its condition.

The President believed those systems could meet much of our needs, at least on a regional level, assuming those systems were in operational condition and could be brought online in short order. This would spare many local utilities the cost of constructing and installing replacement equipment. All the President had to do was get Congress to appropriate the funds to get it back in service. This would give us time to properly design and finance an electrical system for the future, while keeping it affordable. What the President didn't take into account, the lobbyists working for the electrical conglomerates. Those people told Congress and media the industry would lose hundreds of millions of dollars if the President moved ahead with his plan. The President quickly reminded them he had the authority, under the Emergency Powers Act, to do whatever was necessary for the good of the nation, and he intended to do just that whether they liked it or not. For a time Nelsen was the loneliest person in Washington but when he reported his intention to the nation he gathered wide spread support. Unfortunately support can be fickle at times. People still wanted to know when things would be back to normal.

New Realities for Families & Individuals

*L*ife across America had changed dramatically since the September storm, not to minimize the effects of the May event. The hustle and bustle of daily life had slowed considerably. No more chatting with friends via the cell phone, no more instant access via Wi-Fi, no more instant banking, no more instant buying. All the things young people took for granted and older people came to enjoy were taken away. The young had little to fall back on; older folks remembered how things used to be.

The workplace was also greatly affected. To survive companies had to change how they conducted business, unfortunately the first thing most did was layoff pieces of their workforce. That in turn had a ripple effect in the communities where it occurred. Jobs create jobs, yet the opposite also applies. Prior to May 1, unemployment stood at 3.2 percent nine months later it approached 20 percent and rising. Making matters worse, too many Americans had no savings to fall back on. Others scrambled to make ends meet; and worried about the future. They wanted to know where they stood.

From April 2 to April 30, 2021 the President visited every region of the country. He met with community leaders, business owners but mostly talked with everyday people. He wanted to know how they were getting along and how they were coping. Their stories were quite different than those he read about or learned from his advisors. Despite financial setbacks, which for many were huge, most were cautiously optimistic. Business leaders however didn't share that same belief. They felt the government was responsible for what happened; as such was obligated to make everyone whole. Fortunately not everyone in the business community was quite so negative.

On many occasions the President learned about new and inventive ways to improve the nation's communication system, hopefully preventing future disruption. He also learned what some companies were doing to meet their electrical needs. In certain cases, hydrogen was used to power onsite generators and heat generated from production converted to steam and used to condition air--air conditioning. It was a textbook case of American ingenuity at work. In fact the President was so impressed by what he learned he directed the Secretary of Labor and the

Secretary of Commerce to conduct a thorough analysis of company efforts. The President's meetings with American families also afforded him great insight.

As mentioned before, daily life in America had changed a lot since the May and September events, and in other parts of the world as well. At least temporarily, instant everything became instant nothing. Schedules once packed with commitments, now appear blank. Workers went from having little or no spare time to having far too much. Purchases once commonplace were put on hold. It was back to basics. People started talking to one another--face to face, instead of texting or gabbing on the phone. Out of necessity, they had to decide what was important and what wasn't. There was a sense everyone was in the same boat and there was a benefit to working together instead of complaining and otherwise feeling miserable. Parents became more engaged with their kids; together they shared ideas and divided up tasks. They would become better money managers; wasteful spending was out; meals now eaten at home. For the kids, if they wanted to speak with friends they had to visit them.

Most assuredly, they still had their share of problems especially in the early going. The loss of services was a shock to everyone and it would take time to adjust, if that was possible. Before the storm, most families were dual (two) income households. After the September storm, most were down to one in quick order or else one was working part time. Family income in any case was slashed to the bone. Affluence quickly became anything but. Abundance turned into scarcity. That alone was dramatic enough but for those over extended—financially that is—the impact was far greater. They felt they were losing everything; although people are resilient and capable of doing amazing things when the chips are down. Many reinvented themselves, and those were the stories the President heard time and again. Sure there were some who complained or blamed others for their troubles, the government, but never the President.

At nearly every stop folks were amazed the President took the time to meet with them and hear their concerns. It didn't matter if they were Republican, Democrat, Independent, or apolitical, all gave the President high marks. To his credit, the President didn't sugar coat the blight of the country; the nation and the world had changed, in some ways permanently. There was plenty of work to do and that would take time and much patience. He also promised never lied to them—unheard of for a politician. To the best of his ability he would let them know when major developments and breakthroughs were in the offing. Later he told his wife, his trips were extremely rewarding. Upon returning to Washington, he learned of a major setback involving the satellites.

One Step Forward Two Steps Back

Roth personally made the trip to Washington to deliver the news. Initial reprogramming had proceeded as expected. However, once satellites were load tested—i.e. at one hundred twenty percent of design capacity—nearly all reverted to standby status, in less than ten minutes. No one had clue why this happened; if it couldn't be resolved it would prove a major setback. The financial and communication industry were banking on the fact systems would be up and running very shortly. When word got out about this development, the response was immediate. When Roth and Nelsen sat down to discuss the problem in greater detail, the President received an urgent message from Chair of the Federal Reserve and the Stock Exchange President.

The President told Roth he feared the country was on the verge of financial collapse. Measures initiated by his predecessor were only short term fixes; their effectiveness was waning. Roth told the President, programming not equipment was to blame. He also said the hybrid wave might be reaching the satellites too fast; if so, they—the satellites—aren't able to process—the waves-- quick enough; hence they revert to standby. Roth said he understands the predicament the President is in as it relates to the national economy. Early last fall, the economy stalled and gradually declined since—it was now near a breaking point. The President asked Roth to personally oversee reprogramming efforts and report back on the progress. The concerns were clearly etched on their faces; over the last few months both had aged. When the meeting ended, the Labor Secretary was waiting to speak with the President.

The Secretary didn't beat around the bush; he got right to the point. Both large and medium size companies from every part of the country were going to announce major layoffs. Given the circumstances, the layoffs would take effect immediately. As a result, millions of workers would be out of work; combined that with the estimated 30 million already off the job or otherwise working part-time. Federal and state treasuries couldn't possibly process additional claims for unemployment, at least not in a timely manner, let alone funds such expenditures. The Secretary wanted to know what the President was going to do about it. For the first time, the President actually questioned why he wanted the job in the first place. He had few options, and none were good. Still he had faith in Dr. Roth; the man was a genius, if anybody could find a solution to the satellite problem, it was him. He tried hard to reassure the Secretary, but spoke like he doubted what he was saying.

He worried about the country's mood, which in the last few months, quite low; if it soured even more it may mean every man for himself –a revolution no less. It that was the case, there would be no going back. The America as he knew would become a footnote in history and he would be blamed. If pending layoffs weren't enough, the Congress was also in disarray.

In the House, there was no leadership. Since ninety-seven percent of that body was newly elected, continuity was lacking. Close to forty people were now competing for leadership positions, so nothing of real importance was being considered. That body needed to come together and that wasn't happening. In the Senate, things weren't much better. Nearly thirty senators were voted out of office, resulting in a leadership void. Members that weren't up for reelection became suspicious of House members and seemed unwilling to work with them to any degree. The President repeatedly tried to reach out to both body, but had little success. In his view, members were more interested in themselves than the good of the country. He reminded them if they failed to respond he would act on his own by an invoke Executive Privilege by Executive Order; that got their attention. If he was going down, he wasn't going alone, not without a fight; the nation's future was at stake.

Turning the Corner Perhaps

*F*ollowing his visit with the President, Roth quickly rejoined the team, i.e. those involved in reprogramming. He also invited his daughter and son-in-law to join him. Daughter Patty suggested they rethink their approach. She offered two plausible explanations for the problem. One, perhaps there is a line of programming code on the satellites' onboard computer which forces the system into hibernation mode whenever a transmission signal falls outside design parameters; the second involved strength of the signal, which was probably too strong for the computers to handle. After much discussion, the team decided to investigate each---technicians would study the programming code; Roth and the MIT team would revaluate signal transmission.

After six days of nearly nonstop work, technicians found a glitch in the code and made the necessary change. Yet it was decided to hold off on testing until Roth and his team got back to them about the signal question; that would take another two days. Roth and his group concluded signal frequency was probably high, lowering it slightly would not affect performance, assuming the programming glitch was fixed. Everyone was cautiously optimistic as tests were conducted the following day. No one from either team slept the night before.

The atmosphere in the command center was tense; no one said a word yet you could read what everyone was thinking simply by looking at their faces. Five minutes into the test, no reset, ten more minutes, all test satellites still on line, then thirty minutes, then forty-five. With each passing moment you could see people begin to relax. Although, forty-five minutes was one thing, 24/7 365 days a year was quite another. For next ten days the team worked around the clock to test satellite performance; everything worked as they hoped. Only then did Roth feel confident to call the President and give him the good news. He also says would take another two maybe three weeks before every satellite currently owned by the U.S. was back in service. The President decided and Dr. Roth concurred, no public announcement would be made for at least a week until more information was available concerning system performance. The President needed more time before releasing the news—pending layoffs had to be averted.

The President asked his staff to arrange a meeting with every company president involved in the proposed layoffs. Over a one hundred and forty chief executives, from across the country ascended on Washington to confer. They weren't told the purpose of the meeting; no one from the media got invited. Nearly all assumed the President was giving a pep talk. The only people who knew about satellite performance were Roth and his team members—security was that tight. When the President walked in, nearly every executive appeared indifferent by his presence. Only later were remarks given to the media.

*H*is comments to the executives was as follows, "Let me start by saying thank you for coming on such short notice; I know you're extremely busy, as I am. For the past two weeks Dr. Roth and his team of scientists and technicians have been working diligently on getting our satellites back into service. I can report to you now, our full compliment will be operational within a week; two at the most. As of this moment, twenty-six satellites are undergoing extensive testing before we put them back into active service; the remaining units will be placed in service once performance testing is finished." While the

President spoke, the audience looked on in silent disbelieve. Then with little warning, the room erupted in applause; with attendees pounding their fists on the table or standing in splendid amazement. Once the applause was over the President continued with his remarks. "The return to normalcy will be slow and gradual; it will take patience and persistence. There is a chance not everything will go back as it was. We will learn from this experience and hopefully create something better. That's where you folks come in. It's imperative that your employees remain on the job, not only for their benefit, but for your company as well. We've suffered together and so we must heal together. We have to work together for the good of the nation. I therefore appeal to your sense of integrity and ask for your help. Now that's all I have to say." With that, Nelsen promptly left the room, preferring not to answer questions; he knew there would be some he prefers not to answer. More importantly he didn't care to socialize with these people. During the campaign, some of those same folks harshly criticized him for his position. Most of all, the President was confident we were working our way back. However, it was way too early to gloat. As systems were brought back, additional issues would become apparent.

A Few Hiccups on the Road to Recover

One of the first issues that came to light almost immediately involved the "cloud". In the process of reinitializing systems and related networks, it was discovered data located on the cloud, i.e., present on the actual storage devices, was compromised; more specifically the Index Pointers. The actual data was still there but accessing it quickly or at all nearly impossible. Indexes would have to be rebuilt manually, a very labor and time intensive process with no guarantees. It was also learned that financial and internet records were affected the most. The next hurdle involved hundreds of millions cell phones in service prior to the storms. The only process for bringing them back online entailed reinitializing, which amounted to erasing all existing data—contact names, phone numbers, and most other data, except the owner's name and cell number. Erasure was permanent there would be no way to recover anything. It also entailed having every owner to take his or her phone to a specific location on a specific date at a specific time for reinitializing. All of this would take time, not to mention inconvenience, but what choice was there—none.

The road to recovery proved as challenging as the storms themselves. The reality, things would be different; nothing could change that. The rush for speed and faster performance came at a price; not from man himself but nature alone. It forced us to pause and reevaluate what's important and what really isn't. In the course of all the inconvenience and resulting chaos we discovered how to talk to another person, face to face. For many, the experience was rewarding; impersonal got personal. It to, allowed families to reconnect; decisions had to be made, personal contact made the job easier.

Experts say it will take two years, minimum, before things get anywhere close to normal; for the President that was just a guess. Most assuredly, industries will shutter during that time; but hopefully others will come along to take their place. Socially, things might not be that optimistic. Life events can take a generation to absent themselves and forgotten. More than likely people will come to judge technology with a healthily dose of skepticism. They to, will proceed cautiously and less trusting of intuitions. That may affect how companies operate in the future. Guarantees are not absolutes.

With respect to Dr. Roth and his colleagues most simply wanted to return to their regular jobs; though other people outside their work arena

had other ideas. They preferred them to pursue
speaking engagements or take on advisory roles.
Roth the scientist wanted to spread the word but
Roth the father wished a normal life. President
Nelsen wanted him to serve as his science
advisor. The man was torn about what to do.
Nelsen as well was torn, but first he had to mend
fences with Congress; thankfully the Nation's
governors helped him; he was once one of them.

The communications industry, as a whole, had
work to do. If you don't know where you're
headed any road will take you there. The storms
struck them in the heart and they reeled from
their affects. History teaches nothing lasts
forever; change is constant, in nature and the
human experience. Time will tell if the nation
has the wherewithal to get back on track.

To conclude, I offer some personal observations.
In identifying the above, I'm not suggesting
they'll happen or if they do in the manner I've

identified. That's not to say they're implausible; they aren't. There is plenty of scientific evidence to suggest it is the possible. In my years in the public and private sector, I can attest, planning is a hit or miss proposition; more reactive than proactive; the former being the norm. Nevertheless it's absolutely necessary.

Others books by the author: Contradiction, Dark in the Light, Disgrace, First Gentleman, First Gentleman—Death of the President, It Ain't Over Until It Is, Life, Shadows before the Light, First Ten Days

Previously Published as "Reset'

Rochester, Minnesota

August 2017

www.ingramcontent.com/pod-product-compliance
Lightning Source LLC
Chambersburg PA
CBHW071302220526
45468CB00001B/237